IEE Telecommunications Series 14
Series Editors: Prof. J. E. Flood
C. J. Hughes

LAND MOBILE RADIO SYSTEMS

Previous volumes in this series

Volume 1	Telecommunications networks J.E. Flood (Editor)
Volume 2	Principles of telecommunication-traffic engineering D. Bear
Volume 3	Programming electronic switching systems M.T. Hills and S. Kano
Volume 4	Digital transmission systems P. Bylanski and D.G.W. Ingram
Volume 5	Angle modulation: the theory of system assessment J.H. Roberts
Volume 6	Signalling in telecommunications networks S. Welch
Volume 7	Elements of telecommunications economics S.C. Littlechild
Volume 8	Software design for electronic switching systems S. Takemura, H. Kawashima, N. Nakajima Edited by M.T. Hills
Volume 9	Phase noise in signal sources W.P. Robins
Volume 10	Local telecommunications J.M. Griffiths (Editor)
Volume 11	Principles and practice of multi-frequency telegraphy J.D. Ralphs
Volume 12	Spread spectrum in communications R. Skaug and J.F. Hjelmstad
Volume 13	Advanced signal processing D.J. Creasey (Editor)

LAND MOBILE RADIO SYSTEMS

Edited by
R. J. Holbeche

Peter Peregrinus Ltd on behalf of the Institution of Electrical Engineers

Published by: Peter Peregrinus Ltd., London, UK.

While the author and the publishers believe that the information and guidance given in this work is correct, all parties must rely upon their own skill and judgment when making use of it. Neither the author nor the publishers assume any liability to anyone for any loss or damage caused by any error or omission in the work, whether such error or omission is the result of negligence or any other cause. Any and all such liability is disclaimed.

ISBN 0 86341 049 9

Land mobile radio systems.—(IEE telecommunications series; 14)
1. Mobile radio stations
I. Holbeche, R. J.   II. Series
621.3841'65   TK6570.M6

ISBN 0-86341-049-9

Printed in England by Short Run Press Ltd., Exeter

Contents

LIST OF CONTRIBUTORS

Chapters 1 and 7
Dr. R.J. Holbeche
School of Electrical Engineering
Bath University

Chapter 2
Prof. J.D. Parsons
Dept. of Electrical & Electronic
Engineering Liverpool University

Chapter 3
Dr. M.F. Ibrahim
BT Radiophone TSCR
London
and
Dr. A.S. Bajwa
Dept. of Electrical Engineering
Southampton University

Chapter 4
Dr. J.G. Gardiner
Postgraduate School of Electrical and
Electronic Engineering
Bradford
and
Dr. R.E. Fudge
BT Radiophones
95 Ebury-Bridge Road
London SW1

Chapter 5
S.W. Watkinson
Pye Telecommunications Ltd
Cambridge

Chapter 6
Dr. R.C.V. Macario
Dept. of Electrical & Electronic
Engineering
University College of Wales
Swansea

Chapter 8
P.J. Mabey
Phillips Research Laboratories
Redhill

Chapter 9
A.K. Sharpe
Pye Telecommunications Ltd
Cambridge

Chapter 10
R.M. Skiffins
D.T.I.
Directorate of Radio Technology
London

Chapter 11
A.R. Potter
British Telecom
Martlesham Heath

Chapter 12
R.W. Gibson
Phillips Research Laboratories
Redhill

Chapter 13
Dr. J.M. Garrett
Cellnet Centre
British Telecom
London

Chapter 14
Prof. P.G. Farrell
Electrical Engineering Laboratories
Manchester University

Chapter 1
Introduction to mobile radio

R.J. Holbeche

1.1 INTRODUCTION

This book is concerned with the techniques and practices of Civil Land Mobile Radio Communications, generally known simply as Mobile Radio, and regarded merely as the technology of Police and taxi radios. However, the scientific complexity and commercial significance of this important area of activity is considerable, and whereas the commercial significance of this important area of activity is considerable, and whereas the commercial significance of mobile radio is a consideration for companies and governments it is with a view to enhancing the technological and scientific understanding and appreciation of the subject that this book has been prepared by acknowledged experts in the field.

1.2 POLICE AND TAXI RADIOS

Mobile radio is of course concerned with both Polic and taxi radios, identifying samples of two of the major groups of users with significantly different requirements. The public safety, i.e. the Police, Fire and Ambulance authorities have a specific requirement to operate within well defined boundaries the geographic area of which will generally be in excess of that which can be covered by a single base station installation. They will therefore require a multisite scheme to achieve coverage. Also, due to the urgency of their business and the number of users involved, opration on more than one channel is, in addition, a most likely requirement. Rapid access to the Police National Computer (P.N.C.) and a possible requirement for location and status information will also generate a requirement for high speed reliable data communication. Therefore the multisite, multichannel, possibly computer controlled, speech and data communication system will be substantially different from the radio scheme required by the small taxi firms who is most likely to operate over a much smaller area and one that can be covered by a single base station. However, the need for urgency of communication is no less significant than the police authority who, unlike the taxi driver, do not have to survive in a ruthless atmosphere of commercial competition.

The communication requirement for the country vet who will need the facility for efficient and reliable message handling, who unlike the taxi driver does his work away from his vehicle will again be quite different from, say, the ail company senior executive for whom immediate access to the international telephone system may be his most important requirement.

Historically mobile radio equipment has been vehicle mounted. However, because the actual need has generally been to talk to the man and not his vehicle a wide range of portable equipment has emerged. Technological advances in semiconductor devices which due to their small size and modest power requirements are particularly suitable for portable equipment, has led to truly portable radio telephones and pagers. A pager is a small body-worn receiver which can convey a simple message to the user in the form of an alerting tone, a short recorded message or a line of text displayed on the pager itself. These compact, sophisticated equipments embody a large number of significant technological advances over the early vehicle mounted sets with thermionic valves and rotary convertors that were produced in the late 1940's.

1.3 HISTORICALLY

Immediately after World War II, during which radio communication technology advanced dramatically it became possible to produce vehicle mounted radio transmitters and receivers to operate in the VHF band of frequencies which could be used to satisfy the need for improved communications in the police and fire services. At that time the licensing authority which was the Post Office made available a number of channels for private and business use. The very first of those licences were for a fleet of tugs in the Tyne and Wear rivers and Camtax radio taxis in Cambridge both of which were equipped by Pye Telecommunications in 1947.

These early valve equipments were bulky and inefficient, and were vehicle mounted by sheer necessity. A major part of the early technical development of mobile radio did no more than effectively exploit the advances in device technology of the day. Thermionic valves became smaller and more efficient eventually to be replaced entirely by germanium and then silicon transistors. Each quantum improvement in devices producing a smaller, more efficient and often more attractive radio.

This eventual widespread and cost effective availability of reliable transistors combined with the very slow but steady development of battery technology enabled truly portable rather than transportable personal radio telephones to be developed.

The advent of the digital integrated circuit, apparently ideal to produce frequency synthesisers and essential for handling data transmission did not immediately produce a new range of sophisticated equipments as might have been expected. It ook many years and the development of the microprocessor to achieve the next significant advance in mobile radio equipment design. A present day mobile radio will consist of a radio transmitter and receiver, a frequency synthesiser, a data modem and a microprocessor. So that, in theory at least, the radio can perform any desired function to control the system as required. These intelligent radios are necessary in all the various cellular systems.

1.4 RADIO SCHEMES

A basic mobile radio scheme will include a fixed, or base, station, transmitter and receiver located either in a tall building or on a hill top which is connected back to the radio operator, sited usually in an office, either by a dedicated telephone line or often by a pair of low power radio transmitters and receivers connected to highly directional aerials. These radio link equipments were introduced when spectrum was available largely on the grounds of cost, convenience and reliability. They were operated not only on different frequencies but in a different band of frequencies from the base station to reduce the probability of interference.

There are a number of potential sources of interference in mobile radio systems. The most obvious will be when two base stations have been allocated the same frequency but are insufficiently spaced apart so that a mobile radio will experience cochannel interference in its service area. In areas of low mobile radio usage the problem is clearly avoidable, however, in areas of substantial usage the problem becomes largely unavoidable. Intelligent and informed frequency planning, which in practice is a most difficult task, is vitally important to achieve optimum satisfactory performance from a large number of mobile radio schemes operating in the same area.

The output spectrum of any transmitter will contain frequencies other than the main transmit frequency and likewise all radio receivers will respond to frequencies other than that for which they were intended to operate. As the numerical values of these spurious emission and spurious responses will be dependent on the equipment types, avoidance of this problem is beyond the reasonable capability of the planners and when interference occurs can only sensibly be tackled by the equipment users and suppliers.

However, by far the most serious problem of interference in spectrally congested areas is caused by electrical nonlinearity. This occurs principally in the output stages of transmitters and the input stages of receivers although even rusty bolts on transmitter aerial masts can be culprits. Potentially the problem starts when more than one transmitter operates in a band and increases very rapidly when additional transmitters are installed.

In areas of considerable spectral utilisation it is nonlinearity that is the limiting factor on communication and although this happened many years ago with H.F. communication and was partially resolved by adopting an alternative modulation method it appears that this route is unlikely to be followed by V.H.F. and U.H.F. mobile radio.

The problem can be alleviated somewhat by careful frequency planning because the resulting inter-modulation products produced by the nonlinearity can be calculated with reasonable reliability.

1.5 COVERAGE

Initially all radio schemes consisted of a base station transmitter and receiver located on a convenient hill top and a number of mobiles or portables that could clearly only maintain communication in a invariably irregular but reasonably well defined service area. As the radio coverage area from a convenient hill top and the county boundary for a police force, for example, are invariably different, area coverage techniques were developed to extend and tailor the service area by using additional base station sites although still the total service area for any radio scheme rarely exceeds the size of a county.

Much later, as equipment prices reduced and the cost benefit of mobile radio increased dramatically due to the escallation in fuel and vehicle costs, the benefits to be obtained attracted a far wider range of users whose coverage requirements exceeded the defined boundaries of the public safety authorities. Enterprising organisations, calling themselves message handlers, introduced efficient radio schemes combined with a highly professional secretarial service on a virtually national basis.

As the list of mobile radio users expanded rapidly the need was clearly established for a truly national facility when any one could have in his vehicle, or even his pocket, a telephone instrument at least as convenient as the one in his home or on his desk, and preferably more reliable.

The techniques necessary to realise this facility have been embodied in a system which has become known as cellular radio and will eventually satisfy a substantial proportion of the national mobile radio need.

It is interesting to make a comparison between the development of mobile radio systems and the development of the land telephone system, and just as the requirement for private dedicated landlines remains then so will the requirement for private dedicated radio telephone systems.

Chapter 2

Radio wave propagation

J.D. Parsons

2.1. INTRODUCTION

Electromagnetic waves are usually launched into space by energising a suitable antenna. At frequencies below 1 GHz the antenna normally consists of a wire or wires of a suitable length coupled to the transmitter via a suitable transmission line. At these frequencies it is relatively easy to design an assembly of wire radiators which form an array, and beam the radiation in a particular direction. For distances large in comparison with the wavelength and the dimensions of the array, the field strength in free space decreases inversely with the distance, and a plot of the field vs angle is known as the radiation pattern of the antenna.

We can design an antenna to have a radiation pattern which is not omnidirectional, and it is convenient to have a figure of merit to quantify the ability of the antenna to concentrate the radiated energy in a particular direction.

The directivity D of an antenna is defined as

$$D = \frac{\text{Power density at a distance d in the direction of maximum radiation}}{\text{mean power density at distance d}}$$

This is a measure of the extent to which the power density in the direction of maximum radiation exceeds the average power density at the same distance.

The directivity involves knowing the power actually transmitted by the aerial, and this differs from the power supplied at the aerial terminals by the losses in the antenna itself. From the system designers point it is more convenient to work in terms of this quantity and a power gain G is defined as

$$G = \frac{\text{Power density at a distance d in the direction of maximum radiation}}{P_T/4\pi d^2}$$

P_T = power supplied to the antenna

(Note: $G/_D$ is the efficiency of the antenna).

So, given P_T and G it is possible to calculate the power density at any point in the far field that lies in the direction of maximum radiation. A knowledge of the radiation pattern is necessary to determine fields at other points.

The power gain is unity for an isotropic antenna, i.e. one which radiates uniformly in all directions, and so an alternative definition of power gain is therefore the ratio of power density in the direction of maximum radiation to that for an isotropic aerial radiating the same power. We can calculate the power gain of an array by integrating the outward power flow from the array and relating it to the mean power flux. For example, the power gain of a $\lambda/2$ dipole is 1.64 (2.15 dB) in a direction normal to the dipole, and is the same whether the antenna is used for transmission or reception.

The concept of effective area is useful in considering receiving antennas. If an antenna is placed in the field of an electromagnetic wave the received power available at its terminals is the effective area x power per unit area carried by the wave, i.e. P = AW.

It can be shown (see Ref.1, Chap.11) that the effective area of an antenna and its power gain are related by

$$A = \frac{\lambda^2 G}{4\pi}$$

If we consider two antennas separated by a distance d in free space, then if the transmitting antenna has a gain G_T the power density at a distance d is

$$\frac{P_T G_T}{4\pi d^2}$$

So the power available at the receiving antenna, which has an effective area A is

$$P_R = \frac{P_T G_T}{4\pi d^2} A$$

$$= \frac{P_T G_T}{4\pi d^2} \frac{\lambda^2 G_R}{4\pi}$$

So that

$$\frac{P_R}{P_T} = G_T G_R \left(\frac{\lambda}{4\pi d}\right)^2$$

The propagation loss is usually expressed in dB and from this equation we can write

$$L = 10 \log_{10} P_R/P_T$$

$$= 10 \log G_T + 10 \log G_R - 20 \log f - 20 \log d + K$$

and for unity gain antennas we can define a "basic transmission loss" L_b as

$$L_b(\text{dB}) = 32.45 + 20 \log f_{MHz} + 20 \log d_{km}$$

This is an inverse square law with distance, so the received power decreases by 6 dB for every doubling of distance. The equation also shows that the transmission loss increases with the square of the frequency and so high gain antennas are necessary to make up for this loss. The question to be answered is to what extent is this formula modified by the presence of atmospheric effects, the presence of the earth, the effects of trees, buildings, hills etc. which exist in, or in close proximity to, the transmission path.

2.2 PROPAGATION OVER A REFLECTING SURFACE

The situation of two mutually visible antennas over a smooth earth of effective radius r_e is shown in Fig.2.1. Their heights above the earth's surface are h_1 and h_2, and above the tangent plane at the point of reflection the heights are h_1' and h_2'. Assuming h_1 and h_2 are small compared to r_e the reflecting point can be determined as

$$d_1 = \frac{d}{(1+h_2/h_1)} \qquad d_2 = \frac{d}{(1+h_1/h_2)} \qquad \text{where } d = d_1 + d_2$$

The effective antenna heights are

$$h_1' = h_1 - \frac{d_1^2}{2r_e} \qquad h_2' = h_2 - \frac{d_2^2}{2r_e}$$

Substitution of these two equations into the expressions for d_1 and d_2 yields a cubic, one root of which is, to reasonable accuracy

$$d_1 = \frac{d}{1+h_2/h_1}$$

The path length difference between the direct and reflected waves is

$$\Delta R = r_1 + r_2 - r_o$$

$$\simeq \frac{2h_1'h_2'}{d} \quad \text{when } d >> h_1, h_2$$

and the received signal strength is given by

$$= E_d \left[1 + |\rho| \, e^{j\Theta} \right]$$

$$= E_d \left[1 + |\rho| \cos\Theta + j \, |\rho| \sin\Theta \right]$$

The magnitude of this signal is

$$= E_d \left[(1 + |\rho| \cos\Theta)^2 + |\rho|^2 \sin^2\Theta \right]^{\frac{1}{2}}$$

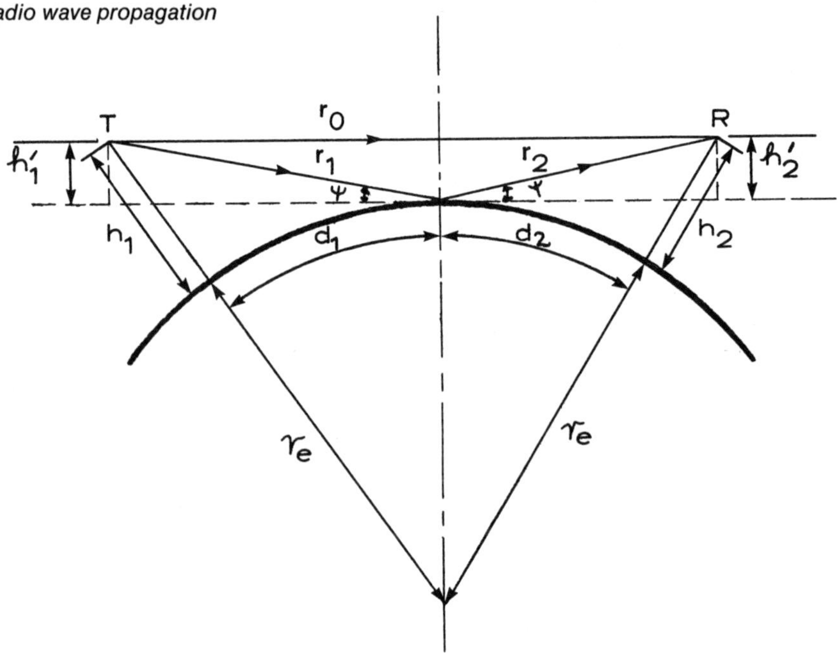

Fig.2.1. Line-of-sight propagation over a spherical earth.

$$= E_d \left[1 + |\rho|^2 + 2|\rho| \cos\left(\frac{2\pi\Delta R}{\lambda} + \phi \right) \right]^{\frac{1}{2}}$$

where ρ is the reflection coefficient of the earth ($\rho = |\rho| e^{j\phi}$) and E_d is the free space field strength.

2.3 PROPAGATION OVER A PLANE EARTH

For a plane earth that is perfectly reflecting $\rho = -1$. So $|\rho| = 1$ and $\phi = -\pi$, and in these circumstances the signal strength is

$$2 E_d \left[1 - \cos \frac{2\pi\Delta R}{\lambda} \right]$$

Substituting for ΔR, and assuming that $d \gg h_1, h_2$ gives

$$P_R = G_T G_R P_T \left(\frac{\lambda}{4\pi d} \right)^2 \left[\frac{4\pi}{\lambda} \frac{h_1 h_2}{d} \right]^2$$

$$\frac{P_R}{P_T} = G_T G_R \left(\frac{h_1 h_2}{d^2} \right)^2$$

In practice, for frequencies above about 100MHz and for horizontal polarisation, $|\rho|$ exceeds 0.9 for angles less than 10° above the horizon over "average" earth. For vertical polarisation $|\rho|$ exceeds 0.5 for angles less than 5°, but must be of the order of 1° or less for $|\rho|$ to exceed 0.9. "Average" earth has a conductivity of about 0.005S and a relative permittivity of about 15.

2.4 THE EFFECT OF SURFACE ROUGHNESS

The surface of the earth is not really smooth, and some criterion is needed to judge whether in any particular situation, the surface will give an appreciable specular reflection. The usual criterion is to define a factor C as

$$C = \frac{\sigma_h \sin\Theta}{\lambda} \quad \left(= \frac{\sigma_h \Theta}{\lambda} \text{ when } \Theta \text{ is small i.e. grazing}\right)$$

where σ_h = standard deviation of the surface irregularities relative to the mean height of the surface.

λ = wavelength

Θ = angle of incidence.

Experimental evidence suggests that if $C < 0.1$, then there is specular reflection and the surface can be considered smooth. For values of $C > 10$ the reflected wave is almost insignificant.

2.5 PROPAGATION OVER TERRAIN OBSTACLES

The preceding discussion has assumed the earth to be a perfectly smooth sphere while in practice the transmission path is modified by the presence of obstructions such as hills, trees and buildings. It is well known that hills and large obstacels along a radio-wave propagation path cause shadows within which large reductions of field strength occur at UHF.

If the terminals of a line-of-sight path are so low that parts of the path pass close to the surface of the earth, the transmission loss due to diffraction effects will be well in excess of the free-space value even if the path is not directly obstructed. When the direct ray clears the obstruction the shadow loss approaches 0 dB in an oscillatory manner as the clearance is increased. Thus a substantial clearance is required over line-of-sight paths in order to obtain "free-space" transmission. Clearly a quantitative measure of clearance (or obstruction) is required. The extent of clearance of terrain features from the direct line-of-sight (or obstruction by terrain features) may be assessed in terms of Fresnel Zone Ellipsoids drawn around the path terminals.

2.5.1. Fresnel Zones

Consider a transmitter T and a receiver R in free space as in Fig.2.2. Imagine a plane perpendicular to TOR at some point along the transmission path. On this plane we sketch concentric circles which represent the loci of the origin of secondary waves travelling to R. The radii of the circles are such that the total path length from T to R via each circle is $n\lambda/2$ greater than that via TOR, where n is an integer. Thus $TA_1 + A_1R$ is $\lambda/2$ longer than TOR, and $SA_2 + A_2R$ is λ longer than TOR etc. The contribution of each successive zone to the field at R is thus alternately positive and negative. These zones are called Fresnel zones and the radius of the n-th Fresnel zone at a point defined by the geometry of Fig.2.2 is

$$R_n = \sqrt{\frac{n\lambda\, d_1 d_2}{d_1 + d_2}}$$

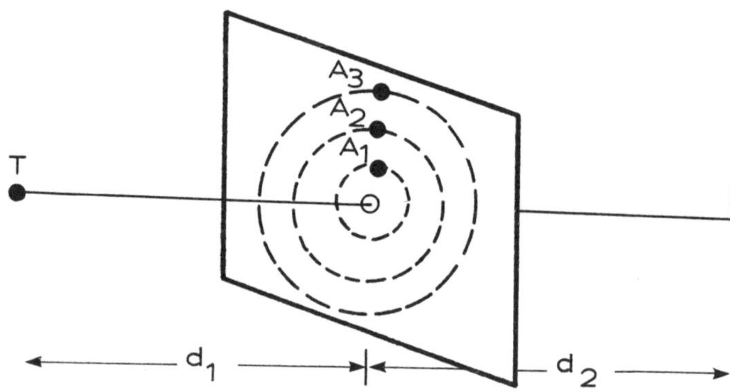

Fig.2.2. Illustrating Fresnel zones

The total energy contributed by all Fresnel zones may be shown to be equal to just half the energy contributed by the central zone of radius OA_1. Thus if an obstructing screen were actually placed at the position shown, with a hole radius OA_1 in the centre, the field strength at R would double. However if the screen let the energy of both first and second zones through, the field at R would be very small, since the contribution of successive zones are almost equal, and 180° out of phase. As the screen aperture is increased so the field at R oscillates, first from zero to twice the final value and back to zero, and then continues to oscillate about the final value with an ever decreasing amplitude of oscillation.

If a straight edge screen (considered as perfectly absorbing) is interposed between T and R as shown in Fig.2.3 then the field at R will begin to oscillate as the height of the screen is increased and the amplitude of the oscillation will reach a maximum when the obstructing edge is just in line with T and R. At this position of the obstructing edge the field strength is just half of the unobstructed value (i.e. the loss is 6 dB). If

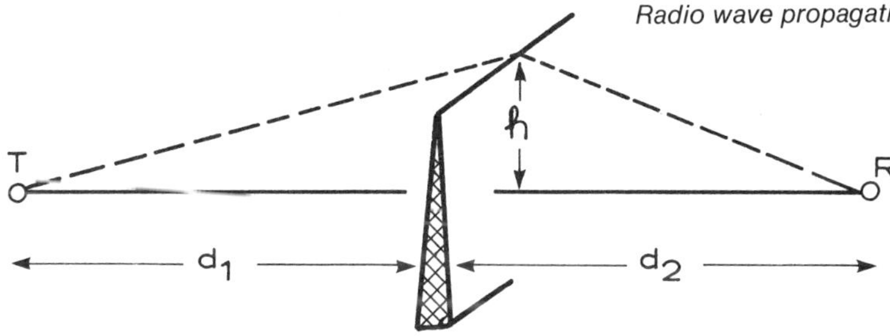

Fig.2.3. Knife-edge diffraction

the height of the screen is greater than this value the oscillation ceases and the field strength decreases steadily to zero.

To express this in a quantitative way we use classical diffraction theory (well documented in many books on optics) extended to radio propagation and we replace any obstruction along the path by an absorbing plane placed at the same position. The plane is normal to the direct path and extends to infinity in all directions except vertically, where it stops at the height of the original obstruction. All ground reflections are ignored.

The signal strength at a point R as shown in Fig.2.3 is determined as the sum of all contributions from the secondary Huygens sources in the plane above the edge. Thus yields an integral of the form

$$F(v) = \frac{(1+j)}{2} \int_v^{\infty} \exp(-j \frac{\pi}{2} t^2)\, dt$$

which is known as the Fresnel integral. v is the Fresnel diffraction parameter given by

$$v = h \sqrt{\frac{2}{\lambda} \frac{(d_1 + d_2)}{d_1 d_2}}$$

A graphical solution of this integral is shown in Fig.2.4 as a function of v. It can be seen that if about 55% of the first zone around a radio path is kept free from obstructions then we avoid the 6 dB loss associated with grazing incidence. In practice designers of point-to-point links try to make the heights of their towers such that as much as possible of the first Fresnel zone is unobstructed, although there is not much to be gained from obtaining additional clearance once the 55% has been achieved.

For $v > 2.4$ the integral may be evaluated by series expansion of the integrand as

$$F(v) = \frac{2}{\sqrt{2}\pi v}$$

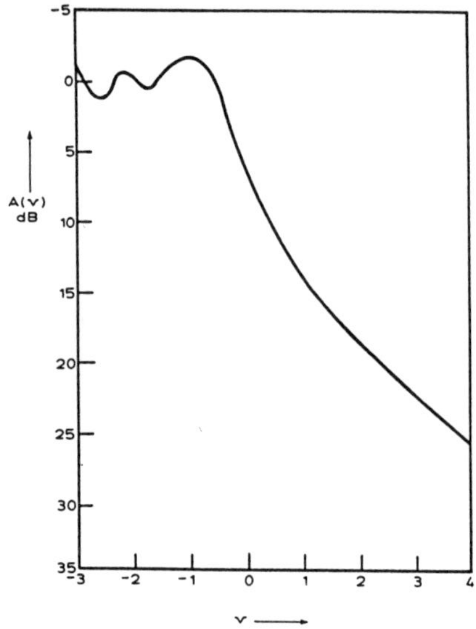

Fig.2.4. Knife-edge diffraction loss

Approximations to the solution are available for $v < 2.4$ as:-

$$-10 \log_{10} | F(v) | = 6.02 + 9.11v - 1.27 v^2 \text{ for } 0 \leq v \leq 2.4$$

$$-10 \log_{10} | F(v) | = 6.02 + 9.0 v + 1.65 v^2 \text{ for } -0.8 \leq v \leq 0$$

2.5.2. Diffraction over a Perfect Knife-Edge with Ground Reflections

If a situation is possible where ground reflections may occur on either side of the knife-edge obstruction of Fig.2.3, a possibility of four geometric paths exists as shown in Fig.2.5. For this reason, it is commonly termed the "four ray" diffraction method.

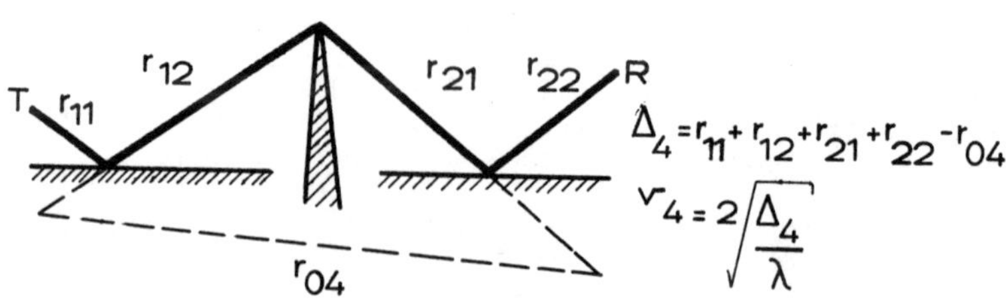

Fig.2.5. Knife-edge diffraction with ground reflections.

It should be noted that the rays have travelled different distances and will be out of phase at the receiver. They also have differing v parameters and, as a consequence, different diffraction losses. The vector sum of all the rays must be taken at the receiver.

$$\frac{E}{E_o} = a(v_1)\, e^{j\phi_1} + a(v_2)\, e^{j\phi_2} + a(v_3)\, e^{j\phi_3} + a(v_4)\, e^{j\phi_4}$$

2.5.3. Diffraction over Rounded Obstacles

Objects which are encountered in the physical world frequently have large thicknesses compared to the wavelength of transmission, e.g. mountains or trees. In these cases, it is necessary to account for effects other than knife-edge diffraction (Ref.2).

The treatment normally given to the problem of a rounded obstacle is to replace it by a conducting cylinder of equal radius to the crest, in a manner similar to that of Section 2.5.1 (Fig.2.6). The cylinder supports reflections on either side of the hypothetical knife-edge which modify the Huygens wavefront above it. The final solution is similar to that of the "four ray" method.

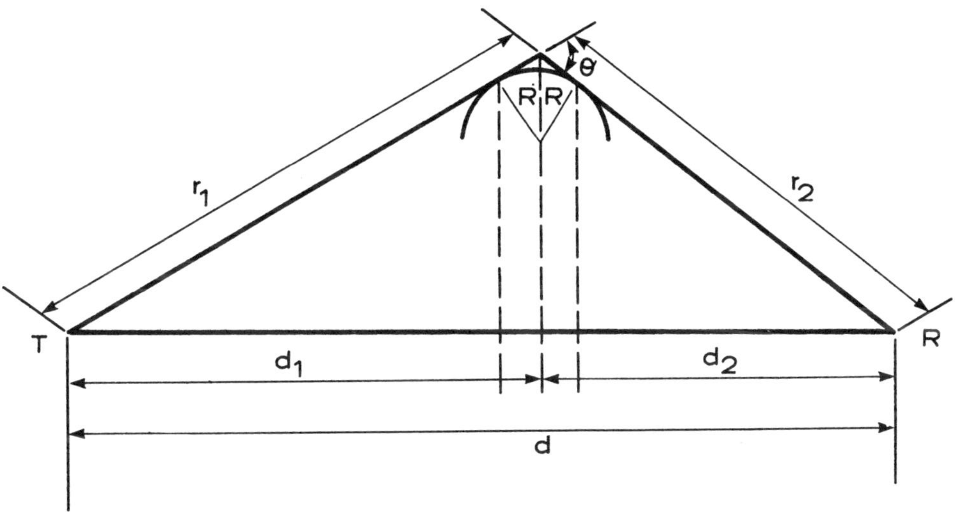

Fig.2.6. Diffraction over a cylinder

An alternative solution is obtained by defining a dimensionless parameter ρ as

$$\rho = \left[\frac{\lambda}{\pi}\right]^{1/6} R^{1/3} \left[\frac{d}{d_1 d_2}\right]^{1/2}$$

The diffraction loss may then be represented by a two dimensional quantity A (v,ρ) expressed in dB. This is related to the one dimensional (perfect knife-edge) loss A (v,O) and a "curvature loss" A (O,ρ) by:-

$$A(v,\rho) = A(v,O) + A(O,\rho) + U(v\rho)$$

where $U(v\rho)$ is a correction factor. The functions $A(v,O)$, $A(O,\rho)$ and $U(v\rho)$ are shown in Figs.2.4, 2.7 and 2.8 respectively.

Approximations for $A(v,O)$ have been given in Section 2.5.1 and for $A(O,\rho)$ and $U(v\rho)$ are

$$A(O,\rho) = 6 + 7.19\rho - 2.02\rho^2 + 3.63\rho^3 - 0.75\rho^4 \quad \text{for } \rho < 1.4$$

$$U(v\rho) = (43.6 + 23.5v\rho)\log_{10}(1 + v\rho) - 6 - 6.7v\rho \quad \text{for } v\rho < 2$$

$$U(v\rho) = 22v\rho - 20\log_{10}(v\rho) - 14.13 \quad \text{for } v\rho \leq 2$$

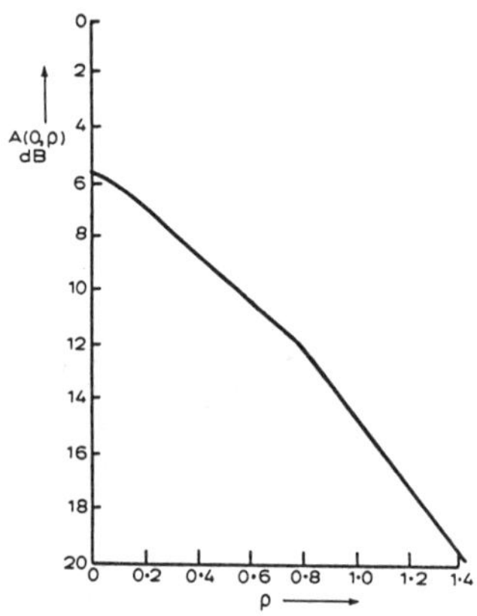

Fig.2.7. Diffraction loss over a cylinder

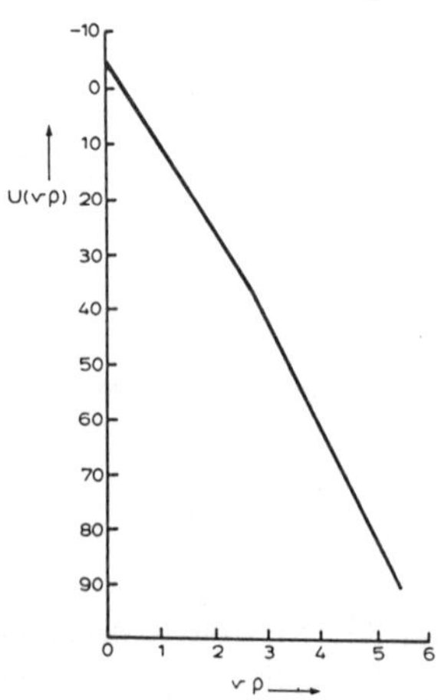

Fig.2.8. Correction factor U for diffraction over a cylinder

The curves of Figs.2.7 and 2.8 are strictly valid for horizontal polarisation only but measurements have shown that at VHF and UHF frequencies they may be applied to vertical polarisation with reasonable accuracy.

Increasing the surface roughness of the cylinder reduces the magnitude of the surface reflections with the result that the solution tends towards that of a perfect knife-edge.

There are a number of ways by which improvements can be made. There are a number of prediction methods based on empirical results and an accurate method of predicting losses due to double knife edge diffractions has been written up in the literature. Various computer techniques for service area prediction for broadcast and land mobile services in the VHF and UHF bands are available, and find wide application. These techniques are capable of predicting the median field strength to within about 6 dB and this is usually considered to be satisfactory.

The effort at present is towards extending these techniques to include urban areas, and to take account of features such as building density, building height, street width and street orientation.

2.6 MULTIPLE KNIFE-EDGE DIFFRACTION

The extension of the single knife-edge diffraction theory to two or more obstacles involves considerable mathematical complexity. The problem reduces to a double integral of the Fresnel form over a plane above each knife-edge and has been solved analytically (Ref.3).

The length and mathematical intricacy of the exact solution has made the use of approximations favourable, especially for more than two edges. The more notable of these are described below.

2.6.1. The Bullington Equivalent Knife-Edge Method

An early proposal for a model which would enable the diffraction loss for a multiply obstructed path to be evaluated was given by Bullington in 1947 (Ref.4). This involves the replacement of the real terrain by a single equivalent knife-edge at the point of intersection of the optical paths made by each terminal and its horizon (Fig.2.9a). The diffraction loss of this equivalent knife-edge is computed as representing the loss of the original real terrain situation.

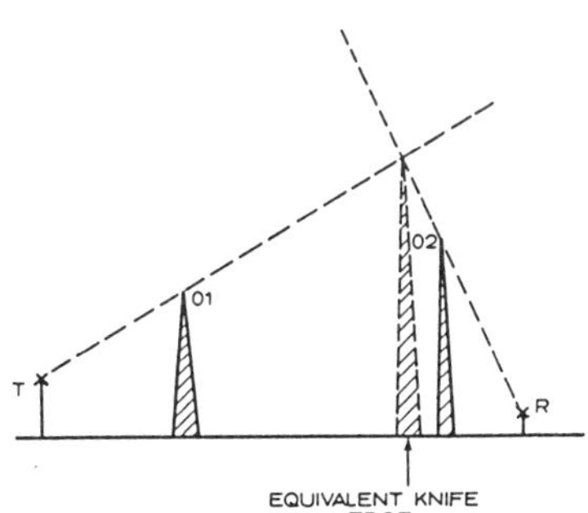

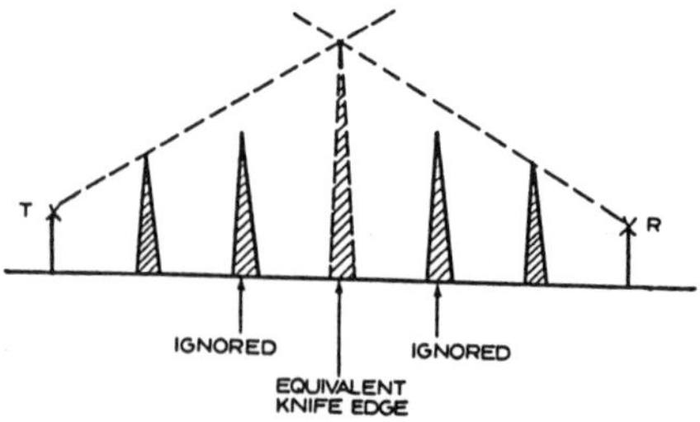

(b)

Fig.2.9. (a) The Bullington equivalent knife edge construction

(b) Equivalent knife edge construction for four obstacles.

Fig.2.9(b) depicts a four obstacle path from which an equivalent knife-edge has been derived. This is completely independent of two of the obstructions and must, therefore, seriously oversimplify the situation when large numbers of obstacles are involved.

2.6.2. The Epstein-Peterson Diffraction Loss Method

The Epstein-Peterson diffraction loss method (Ref.5) calculates the propagation loss of a multiply obstructed path by adding the attenuations (in dB relative to free space) produced by each knife-edge in turn. A two obstacle situation depicted in Fig.2.10 is evaluated as the sum of the diffraction losses for paths T-01-02 and 01-02-R.

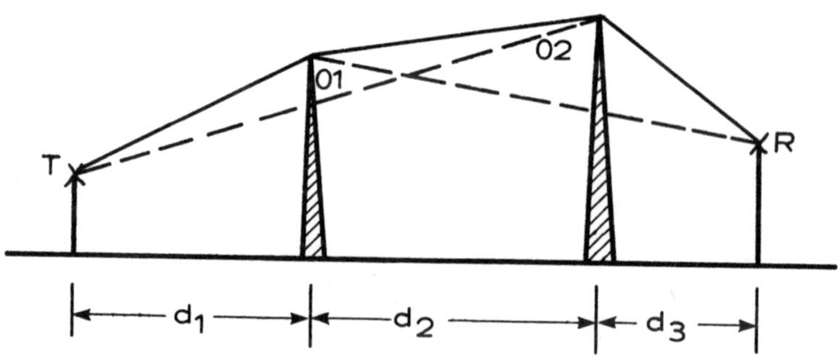

Fig.2.10. The Epstein-Peterson diffraction construction

Comparison of this method with the rigorous solution for two knife-edges has revealed that large errors exist when the two obstacles are closely spaced.

2.6.3. The Japanese Atlas Method

This method, proposed by the Japanese postal service (Ref.6) is shown in Fig.2.11 and is similar to the Epstein-Peterson method. The difference is that the diffraction loss is computed for paths T-01-02 and T'-02-R where T' is the projection of the horizon ray 01-02 onto the plane of T.

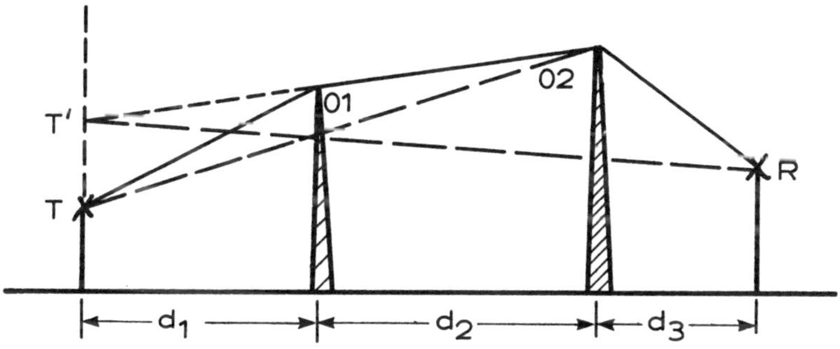

Fig.2.11. The Japanese Atlas diffraction construction

Inclusion of a correction factor to the Epstein-Peterson method for cases of two closely-spaced edges has shown that it then becomes identical to this method. The comments on accuracy made in the previous section therefore hold.

2.6.4. The Deygout Method

The Deygout method (Ref.7) of determining the diffraction loss over obstructed paths, often called the 'main-edge' method, is best described by way of an example.

Consider the three obstacle paths of Fig.2.12. The losses due to each of the knife-edges must be evaluated in the absence of the other, i.e. paths T-01-R, T-02-R and T-03-R. The larger is termed the 'main-edge' and its diffraction loss calculated (path T-02-R in Fig.2.12). The diffraction loss over the remaining obstacles is then found with respect to the main-edge and the visible terminal (T-01-02 and 02-03-R).

For more than three knife-edges, the total loss is evaluated as the sum of all the individual losses for the edges in order of decreasing loss when the above procedure is recursively repeated. For a double knife-edge case in which one edge is largely dominant good agreement with the exact solution is obtained. When the magnitudes of the losses are comparable however the method is found to overestimate the loss.

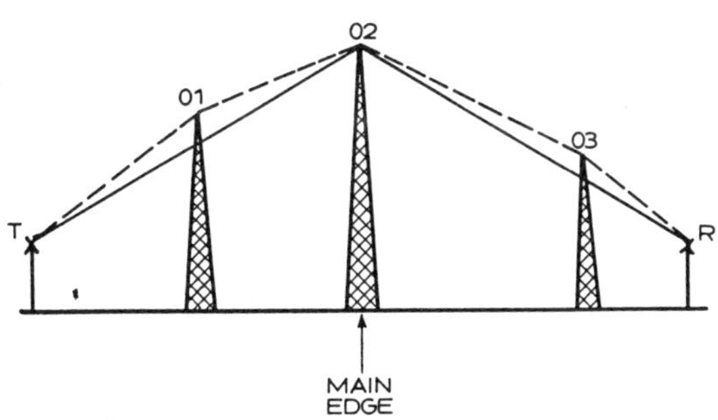

Fig.2.12. The Deygout diffraction construction

Comparison of the predicted loss (without corrections) for three and more knife-edges with measurements at 160, 1800 and 2450MHz shows close agreement for highly obstructed paths. Superior accuracy to the Bullington and Epstein-Peterson methods has also been demonstrated but the application to paths with more than four obstacles results in a complex solution which generally overestimates the losses.

2.6.5. Comparison of the Diffraction Models

Comparisons of the diffraction loss approximations described in the previous sections have been given in the literature. Each model has been found to have a region over which it is most suitable but, as a general rule, the Epstein-Peterson construction produced consistently low errors when compared to the rigorous two obstacle solution.

Over highly irregular terrain the Deygout method has been demonstrated to give increased accuracy. The improvement generally appears small, however, and may not warrant the additional complexity involved.

2.7. PROPAGATION PREDICTION MODELS

As an indication of the manner in which transmission loss calculations may be made, an examination of the more notable irregular terrain prediction models will be given. The basic elements of these are to be found in the previous sections. The applicability of each model, where defined by the authors, is given in Table 1.

2.7.1. The Egli Method

This semi-empirical prediction model was developed in 1957 after an exhaustive series of measurements over irregular terrain (Ref.8). The basic method forces agreement between the plane earth loss (Section 2.3) and the measured value by use of a correction factor as a function of frequency, aerial heights and polarisation.

The important point of the model is that the standard deviation of the correction factor is related to that of the terrain undulations. The latter are presumed to be a log-normally distributed random quantity.

2.7.2. The Edwards and Durkin Method

The Edwards and Durkin prediction model (Ref.9) requires that a terrain profile of the transmission path be obtained and used to calculate the diffraction loss over the path. This is subsequently added to the greater of the free space and plane earth propagation losses.

The free space loss normally has the larger value at short ranges in the region where the direct and ground reflected waves are interfering.

Determination of the plane earth loss requires knowledge of the antenna heights. In the original presentation measured values were compared with predictions using the real heights above ground. Close agreement was found to exist at 86.2MHz.

2.7.3. The Blomquist and Ladell Methods

Two prediction models have been proposed (Ref.10) which attempt to describe the transmission path loss. Both methods define the following loss factors as:-

L_{fp} : Plane earth transmission loss in excess of free space

L_{fd} : Diffraction loss over the transmission path (in excess of free space).

L_f : Free space transmission loss

Method (1) Total Loss = L_f + the larger of L_{fd} and L_{fp}

Method (2) Total Loss = $L_f + (L_{fd}^2 + L_{fp}^2)^{\frac{1}{2}}$

When $L_{fd} >> L_{fp}$ (highly obstructed paths), Loss $\simeq L_f + L_{fd}$

When $L_{fd} << L_{fp}$ (unobstructed paths), Loss $\simeq L_f + L_{fp}$ = plane earth path loss.

Method (2) may be seen to smooth out the apparent discontinuity of method (1) but it should be noted that no theoretical justification of either model has been given.

Comparison between the predicted and measured values over highly obstructed paths has revealed that while both give small errors, method (2) is consistently better. Method (1) has, therefore, been discarded in favour of method (2).

2.7.4. The BBC Method

The British Broadcasting Corporation have formulated a model to predict UHF television coverage areas (Ref.11). The assumptions made by this method are that the receiving aerial is at roof-top level (10m) and the transmitting aerial is well clear of local obstructions. The path loss is calculated as the sum (in dB) of the free space path loss, the terrain diffraction loss and an "absorption loss".

The absorption loss term introduced above is intended to account for the building and tree clutter in the vicinity of the receiver. One mile of the terrain profile closest to the receiver is divided into tenths and the absorption loss calculated as a function of building and tree density. A weighting factor is employed to place the largest emphasis on the 0.1 mile section of path closest to the receiver, the next largest emphasis on the next 0.1 mile and so on for all ten segments.

This method requires an extremely detailed data base if predictions within a city are to be produced. The type of information required, however, is not available from Ordnance Survey maps and would require a specialised survey. The commercial applicability would be limited to point-to-point applications.

Application of the model down in frequency to VHF requires that ground reflections be included; extension for use with aerials lower than 10m would need to be the subject of further measurements.

2.7.5. The Okumura Method

A large number of measurements were made over a wide range of frequencies in Japan (Ref.12). The paths were considered to be representative of commonly occurring terrain situations, especially major centres of population. The results were processed to produce an empirical model based on the free space propagation equation and a series of "correction" curves taking account of parameters such as range, frequency, terminal heights and terrain roughness. Precisely because of the areas in which data was taken, it is a model intended for use with radio links operating with one terminal in the vicinity of buildings. Suitable corrections factors are given however, for the cases when the mobile is in suburban and open areas.

Precise relationships will not be given here, but the parameters which must be evaluated to determine the various correction factors are defined below.

(a) The Effective Base Station Height (h_{te}):
The height of the base station aerial above the average ground level calculated over the range interval of 3 to 15km (or less if the total range is below 15km).

(b) The Interdecile Terrain Undulation Height (Δh):
The difference between the 90 and 10 percentiles of the "peak-to-peak" value of the terrain undulation height, computed over the 10km of the path closest to the vehicular station.

(c) Average Ground Slope (Θ_m):
The average ground slope is determined over the 5 to 10km of the terrain path closest to the vehicular terminal.

The following corrections are given in graphical form within the bounds of the relevant variables as specified.

(1) Basic Median Attenuation (A_{mu}):
1 to 100km link range
100 to 3000MHz transmission frequency.

(2) Base Station Height Gain (H_{tu}):
1 to 100km link range
20 to 1000m effective base station aerial height.

(3) Vehicular Station Height Gain (H_{ru}):
1 to 10m vehicular aerial height
100 to 1000MHz for a large city
or 100 to 2000MHz for a medium city.

(4) Rolling Hilly Correction Factor (K_H):
10 to 500m Interdecile terrain undulation height
450 to 1500MHz transmission frequency.

(5) Sloping Ground Correction Factor (K_{SP}):
-20 milliradians to +20 milliradians average ground slope.

Further corrections are available for the situation where the vehicle is free from all obstructions for a range of 300 to 400m (open ground) at frequencies in the range 100 to 2000MHz and in a village or sparsely tree lined road (suburban area) at frequencies from 100 to 3000MHz.

The total path loss is predicted as the free space loss plus (or minus depending upon the context) all the various correction factors.

2.8 COMPARISON OF PREDICTION MODELS

The propagation prediction techniques described previously tend to fall broadly into two groups. These are the classes of free space and plane earth based models.

The simpler methods detailed use the plane earth model independently of frequency and the surface roughness whereas, in reality, the magnitude of the coherent ground reflected ray diminishes as the surface irregularities increase. The effective surface roughness, in turn, increases with decreasing wavelength and increasing grazing angle. It is reasonable to suppose, therefore, that two models are possible:-

(i) Free Space Based at high frequencies (>300MHz say) or with large surface roughness.

(ii) Plane Earth Based at low frequencies (<300MHz say) and with small roughness.

The location of the division between the two models would need careful investigation to establish the criteria governing the applicability of each method.

MODEL	FREQUENCY (MHz)	LINK RANGE (km)
Egli	40 - 900	<60
Edwards and Durkin	30 - 300	<40
Blomquist and Ladell	30 - 900	5 - 22
BBC	UHF (not specified)	<20
Okumura	100 - 3000	10 - 100

Table 2.1. Applicability of the Various Prediction Models

References

1) Jordan E.C. and Balmain K.G. 1968, 'Electromagnetic waves in radiating systems, Prentice-Hall (2nd edn).

2) Hacking K. 'Propagation over rounded hills' 1968 BBC Research Report RA-21

3) Millington G, Hewitt R and Immirzi F.S. 'Double Knife-edge diffraction in field strength predictions' 1962, Proc IEE, 109C, 419-429.

4) Bullington K 'Radio Propagation at frequencies above 30Mc/s' 1947, Proc IRE, 35, 1122-1136.

5) Epstein J and Peterson D.W. 'An experimental study of wave propagation at 850MC' 1953, Proc IRE, 41, 5, 595-611.

6) 'Atlas of radio wave propagation curves for frequencies between 30 and 10000Mc/s' 1957, Radio Research Laboratory, Ministry of Postal Services, Tokyo, Japan, 172-179.

7) Deygout J. 'Multiple Knife-edge diffraction of microwaves' 1966, IEEE Trans, AP-14, 4, 480-489.

8) Egli J.J. 'Radio propagation above 40Mc over irregular terrain' 1957, Proc IRE, 1383-1391.

9) Edwards R and Durkin J 'Computer prediction of service areas for VHF mobile radio networks 1969, Proc IEE, 116, 9, 1493-1500.

10) Blomquist A and Ladell L 'Prediction and calculation of transmission loss in different types of terrain' 1974 NATO-AGARD Conference Publication CP 144.

11) Causebrook J.H. 1974, Computer prediction of UHF broadcast service areas' BBC Research Report RD 1974-4.

12) Okumura Y, Ohmori E, Kawano T and Fuduka K. 'Field strength and its variability in VHF and UHF land mobile service' 1968, Rev. Elec. Commun. Lab, 16, 825-873.

Chapter 3

Propagation in urban areas

M.F. Ibrahim and A.S. Bajwa

3.1 INTRODUCTION

Random fluctuations of the signal strength are a common occurrence in transmissions from a base station to a mobile. These fluctuations arise from the fact that, in a typical urban environment, the line-of-sight propagation path between the transmitter and receiver is usually shadowed by the surrounding buildings and obstacles. Thus radio-waves arrive at the mobile via many indirect paths, each path resulting from reflection or diffraction around the buildings and terrain (Fig. 3.1). These radio-waves combine to set up a spatially-fluctuating standing-wave field. It is the movement of the mobile through such a field that causes the receiver to sense a time-related fading signal. To examine further the mechanism leading to fading, let us consider the transmission of an unmodulated carrier. Each path introduces a phase shift corresponding to its electrical length from the base station to the mobile. The existence of many propagation paths i.e. multipath, means that a number of randomly phase-shifted carrier components appear at the receiver. Depending on the relative phase-shifts, these multipath carriers combine constructively or destructively with the resultant envelope fading as the path lengths change for the different spatial locations. A typical experimental record (Fig. 3.2) illustrates this characteristic. Fades of a depth less than 20 dB are frequent, but the deeper fades i.e. in excess of 30 dB, although less frequent, are not uncommon. Rapid fading is usually observed over short distances of half a wavelength or so.

The characteristics of the multipath change as the mobile moves from one homogeneous site to another, different terrain and overall building features around the mobile change the 'local' phenomenon. As a result, the rapid fades are superimposed on a signal of slowly variable level. The slow variations in the 'local mean' of the fading signal vary according to the gross features in the nature of the terrain and buildings. Sometimes it is also loosely associated with the 'shadow effects' of the urban environment.

If we now consider the nature of the multipath and the shadow effects and their influence on the characteristics

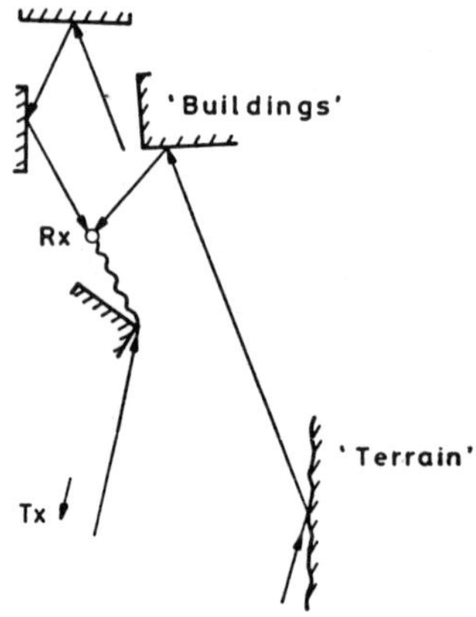

Fig. 3.1 Multipath Scattering Geometry for a Hypothetical Location

of radio-wave propagation, it is apparent that an 'exact' deterministic characterisation is difficult and improper. Thus a resort to the powerful tools of statistical communication theory is inevitable.

In this tutorial review a convenient distinction is drawn between the dichotomy of the short-term multipath effects and the long-term variations of the local means. The multipath models and the statistical characterisation of the short-term, i.e. rapid fading, are considered first. This is followed by a discussion on the long-term variations of the signal and the statistical variability of the local mean. Factors that affect the local means are studied in order to provide suitable prediction models for the median propagation losses. These prediction models answer some of the most relevant questions relating to radio coverage in urban areas.

3.2 SHORT-TERM FADING

Several multipath propagation models have been suggested in order to explain the statistical characteristics of the fields and the associated fading signal envelope. Ossana (1) was the first to attempt an explanation based on interfering waves incident and reflected from the flat sides of randomly placed buildings. Ossana's model predicted theoretical power spectra in good agreement with measured spectra in suburban areas. Evidence suggests (2) that the model is inappropriate, if not inflexible, for urban areas since it assumes the existence of a direct path and is confined to a restricted range of reflection angles. A scattering model is a more appropriate and general model to

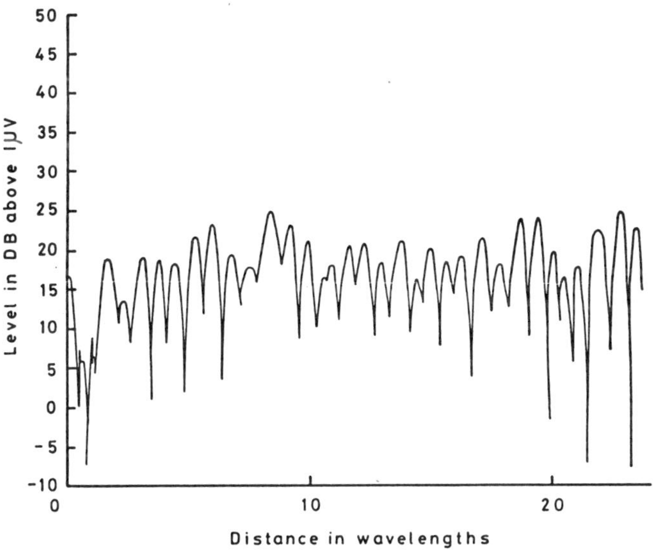

Fig. 3.2 Typical Received Signal, London, 168 MHz

study the mechanisms leading to the fading characteristics.

3.2.1 The Scattering Model

The model presented here was proposed by Clarke (2). It assumes that the field incident on the mobile antenna is a 'scattered field' composed of a number of randomly-phased plane waves. Clarke assumed that the plane waves are vertically polarised with azimuthal angles of arrival and the r.f. phase angles as random and statistically independent. Furthermore, the r.f. phase angles are assumed to have a uniform probability density function p.d.f., over 0 to 2π radians. This is plausible at VHF and above where the wavelength is small enough to ensure that small changes in path length result in appreciable change of the r.f. phase. The p.d.f. $p(\alpha)$ for the azimuthal angle of the plane waves is not specified as it depends on the topography and scattering characteristics of the environment, and is likely to change over spatial distances greater than a few tens of wavelengths. A statistical model, such as this, based on scattered waves allows important relationships to be established for the received signals in the form of first and second order statistics of the signal envelope and to predict the nature of the frequency spectrum. Other scattering models have also been proposed (3,4) and reduce to the same form when the number of constituent waves is large.

At every spatial point exactly N plane waves of equal amplitude are assumed to arrive. This is a valid assumption in heavily built-up areas since the scattered components are likely to experience similar 'path' attenuation. In certain situations a direct line-of-sight path may contribute a steady 'coherent' non-random component. This no doubt will alter the nature of the fading envelope and its statistics, but first let us consider N equi-amplitude waves. The path angle geometry for the nth scattered plane wave is shown in Fig. 3.3. If the transmitted signal is vertically polarised, i.e. the electric field vector is aligned along the z axis, the field components at the mobile are:

i) electric field E_z
ii) magnetic field component H_x
iii) magnetic field component H_y

The expressions for the resultant fields at the spatial point P in Fig. 3.3 are obtained in the complex equivalent baseband form (2) as

$$E_z = E_o \sum_{n=1}^{N} \exp(j\phi_n) \tag{3.1}$$

$$H_x = \frac{-E_o}{\eta} \sum_{n=1}^{N} \sin\alpha_n \exp(j\phi_n) \tag{3.2}$$

$$H_y = \frac{E_o}{\eta} \sum_{n=1}^{N} \cos\alpha_n \exp(j\phi_n) \tag{3.3}$$

In the field equations ϕ_n is the phase relative to the carrier phase, E_o is the real amplitude of the N plane waves, and η is the intrinsic wave impedance. The real and imaginary parts of each field component constitute the in-phase and quadrature components of the received r.f. fields for an unmodulated-carrier excitation at the transmitter.

Applying the central limit theorem and observing that

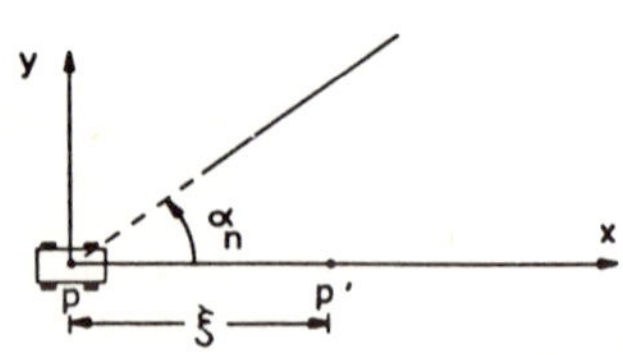

Fig. 3.3 A Component Wave Incident on Mobile Antenna

α_n and ϕ_n are independent it follows that E_z, H_x and H_y are complex Gaussian random variables for large N (5).

If in addition to the scattered wave, a direct wave with significant magnitude reaches the receiver, (a situation which may arise if the vehicle is near the base station or in a rather open area), Rice's analysis of "sine wave plus random noise" may be used (5). The relationship between Rician, Rayleigh and gaussian distributions is illustrated in Fig. 3.4.

The scattering model based on the assumption that the mobile received signal is of the scattered type with each component wave being independent, randomly phased and of random arrival angle, leads to the conclusion that the probability density function of the envelope is Rayleigh i.e.

$$p_r(r) = \frac{r}{b_o} \exp\left(-\frac{r^2}{2b_o}\right) \tag{3.4}$$

and the corresponding cumulative distribution function:

$$P_r(r) = 1-\exp\left(-\frac{r^2}{2b_o}\right) \tag{3.5}$$

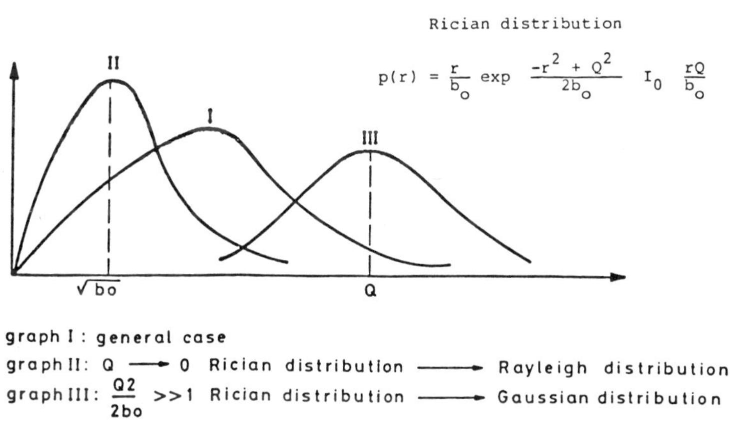

Fig. 3.4 Probability Distributions of the Fading Envelope

The distribution has the following characteristics

$$\langle r \rangle = \sqrt{\frac{\pi b_o}{2}}$$

$$\langle r^2 \rangle = 2b_o$$

3.2.2 Angle of Arrival and Signal Spectra

Whenever the receiver or transmitter is in motion, the received r.f. signal experiences a Doppler shift, the frequency shift being related to the cosine of the spatial angle between the direction of arrival of the wave and the direction of motion of the vehicle. A mobile travelling at a constant speed vm/s along the x-axis in Fig. 3.3 introduces a Doppler shift f_n for the nth plane wave according to the relationship:

$$f_n = \frac{v}{\lambda} \cos\alpha_n \tag{3.6}$$

where λ is the carrier wavelength.
It can be seen that waves arriving from ahead of the vehicle experience a positive Doppler shift, with a maximum value v/λ, and waves from behind the vehicle do so with a negative Doppler shift.

As the number of waves N becomes large, the fraction of incident power in an angle between α and $\alpha + d\alpha$ for an isotropic antenna is $p(\alpha)\, d\alpha$. An antenna with a power gain $G(\alpha)$ and correct polarisation will contribute the fractional power $G(\alpha)\, p(\alpha)\, d\alpha$ in the same incremental angle. By equating this incremental power to the incremental power determined by the r.f. Doppler power spectrum density $S(f)$, we obtain the relationship:

$$S(f)|df| = \{G(\alpha)\, p(\alpha) + G(-\alpha)\, p(-a)\}|d\alpha| \tag{3.7}$$

Note that the r.f. Doppler frequency shift $f(\alpha)$ is related to the angle of arrival of waves α and the carrier frequency, f_c;

$$f(\alpha) = f_m \cos\alpha + f_c \tag{3.8}$$

where $f_m = v/\lambda$.

From Eqn. (3.7) since $f(\alpha) = f(-\alpha)$, we have

$$|df| = f_m|-\sin\alpha d\alpha| = \sqrt{f_m^{\,2} - (f - f_c)^2}\ .\ |d\alpha| \tag{3.9}$$

and substituting in Eqn. (3.7):

$$S(f) = \frac{\{p(\alpha)\, G(\alpha) + p(-\alpha)\, G(-\alpha)\}}{f_m\{\sqrt{1 - \left(\frac{f - f_c}{f_m}\right)^2}\}} \tag{3.10}$$

The Doppler power spectral density S(f) depends on the antenna gain and the angular distribution for the multipath waves. This power spectrum is different for each field component. For example, the electric field sensed by a vertical $\lambda/4$ monopole with $G(\alpha) = 1.5$ is

$$S_{E_z}(f) = \frac{1.5\,\{p(\alpha) + p(-\alpha)\}}{f_m\left\{\sqrt{1 - \left(\frac{f - f_c}{f_m}\right)^2}\right\}} \tag{3.11}$$

Likewise loop antennas may be used to sense the magnetic fields. A vertical loop aligned in a plane perpendicular to vehicle motion, i.e. x-axis, has an antenna gain pattern

$$G(\alpha) = 1.5 \sin^2\alpha \tag{3.12}$$

and aligned along the y-axis, it has

$$G(\alpha) = 1.5 \cos^2\alpha \tag{3.13}$$

By substituting Equations (3.12) and (3.13) in equation (3.10) one can obtain the expressions for the power spectral density assuming an unmodulated carrier transmission.

The practical case of interest is the omni-directional vertical monopole. We can obtain an expression for S(f) by assuming $p(\alpha)$ is uniformly distributed in the angular range $-\pi$ to π. This need not be the case as any other suitable function can be considered. In fact, experiemntal data (6,7) confirms that such an assumption is simplistic since the scattering changes from one location to another even within a homogeneous locality (6). With the simplifying assumption, the Doppler spectrum becomes:

$$S_{E_z}(f) = \frac{1.5}{\pi f_m}\left[1 - \left(\frac{f - f_c}{f_m}\right)^2\right]^{-\frac{1}{2}} \tag{3.14}$$

In the time-domain the effects of the randomly-phased and Doppler-shifted multipath signals appear in the form of a fading envelope. In Refs. (2,3) expressions have been derived for the baseband envelope spectral densities to demonstrate the influence of a direct wave on the power spectrum. The r.f. Doppler spectra and the envelope spectra for $p(\alpha) = \frac{1}{2\pi}$ and the case with a direct wave at angle α_o are illustrated in Fig. 3.5. Note that the envelope spectra has two additional peaks at $f_m(1 \pm \cos\alpha_o)$ when a direct wave is present.

3.2.3 Spatial Correlation of Field Components

In radio systems, especially at VHF and above, there is an advantage in using two or more spatially separated antennas if the signals at each antenna fade independently. Statistically this ensures that the combined signals are

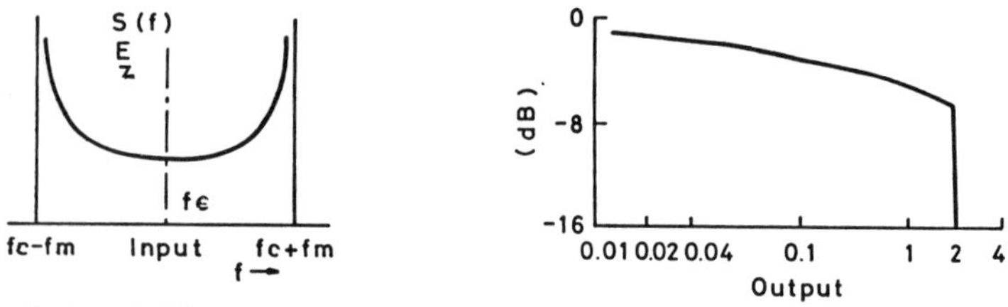

a) Input RF spectrum and baseband spectrum of output envelope for isotropic scattering.

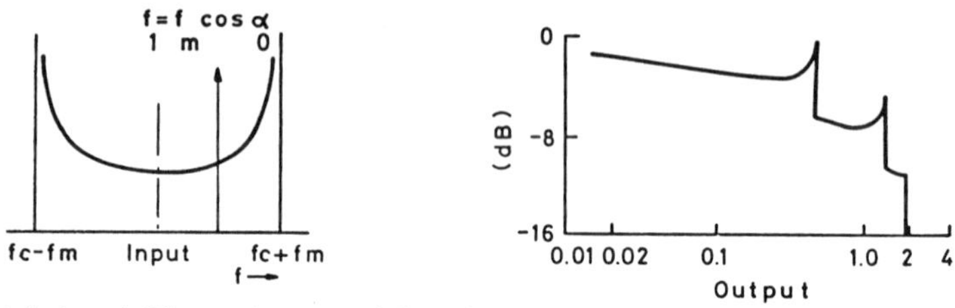

b) Input RF spectrum and baseband spectrum of output for isotropic scattering plus direct wave component with angle of arrival α_0.

Fig. 3.5 Power Spectra for Vertical Monopole

likely to remain at a reasonable level. This technique is known as space diversity.

Let us obtain the spatial correlation function for the electric field. In Fig. 3.3 consider the field incident at P', at a distance ξ along the x-axis from the point P. The phase of the nth component wave at P' will be $\phi_n + k\xi\cos\alpha_n$, where $k = \frac{2\pi}{\lambda}$ is the phase-shift factor in free-space. The auto-covariance function of the electric field is given in the form

$$R_{E_z}(\xi) = \langle E_z^* E_z' \rangle = \langle E_o^2 \sum_{n=1}^{N} \exp(-j\phi_n) \sum_{n=1}^{N} \exp\{j(\phi_n + k\xi\cos\alpha_n)\}\rangle \quad (3.15)$$

where E_z^* is the complex conjugate of the field component at P and E_z' is the field component P'. If we assume statistical independence of ϕ and α, and also $p(\alpha) = \frac{1}{2\pi}$ it follows that (2),

$$R_{E_z}(\xi) = N E_o \int_{-\pi}^{\pi} p(\alpha) \exp\{jk\xi\cos\alpha\} d\alpha = N E_o^2 J_o(k\xi) \quad (3.16)$$

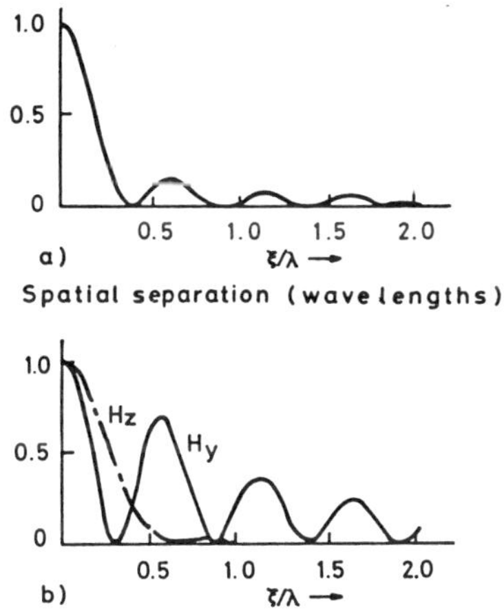

Fig. 3.6 a) Normalised Autocovariance Function of $|E_z|$

b) Normalised Autocovariance Functions of $|H_x|$ and $|H_y|$

The normalised autocovariance function of the envelope $\rho_{|E_z|}$ is of interest rather than the complex field E_z. In Refs [2,3] it is shown that the normalised autocovariance of the envelope is to good approximation equal to the square of the normalised autocovariance function of the complex random field. Thus,

$$\rho_{|E_z|} \cong J_o^{\,2}\,(k\xi) \tag{3.17}$$

where $J_o(.)$ is the zero-order Bessel function of the first kind.

The expressions for the magnetic field components can be similarly obtained. The normalised autocovariance functions for the respective envelopes are plotted in Fig. 3.6. Observe that the scattered fields decorrelate rapidly to a small value within a distance of $\lambda/2$. Therefore, it is practical to employ space diversity even on a mobile.

3.2.4 Fading Envelope Statistics

The fading envelope directly affects the performance of AM receivers. It is of interest to consider the 'rate' at which the envelope crosses a given level and how long it remains below that level. The Rayleigh fading envelope only occasionally experiences the very deep fades, i.e. 30 dB fades occur for less than 0.1 percent of the time. For a specified signal level, R, the level crossing rate N(R) for a crossing in the positive direction is given by (3,4)

$$N(R) = \int_0^\infty \dot{r} \cdot p(R,\dot{r})\, d\dot{r} \tag{3.18}$$

where $p(R,\dot{r})$ is the joint p.d.f. of R and $\dot{r}$, and the dot indicates the time derivative. The joint p.d.f. $p(r,\dot{r})$ is obtained in the form (3).

$$p(r,\dot{r}) = \frac{r}{b_o} \exp \frac{-r^2}{2b_o} \cdot \frac{1}{2\pi\, b_2} \exp \frac{-\dot{r}^2}{2b_2} \tag{3.19}$$

where b_o and b_2 are the moments of the r.f. spectrum [see Ref.(3), p.25] as defined by Rice.

For a vertical monopole the level crossing rate becomes

$$N(R) = 2\pi\, f_m\, \rho \exp\{-\rho^2\} \tag{3.20}$$

where $\rho = \frac{R}{R_{rms}}$ is the ratio of the specified level to the r.m.s. amplitude of the fading envelope. The normalised level crossing rates for the vertical monopole is plotted in Fig. 3.7a. The maximum level crossings occur at ρ = -3 dB and decreases as the level is lowered or increased. Recall that the crossing rate is specified for a positive direction and so when a low level is set the signal remains above this level most of the time. The same applies when a level much greater than the r.m.s. value is selected. At f_c = 900 MHz and a vehicle speed of 48 kph (30 mph), f_m = 40 Hz; thus the level crossing rate N_R is 39 per second at ρ = -3 dB.

The average duration of fades below the level R is

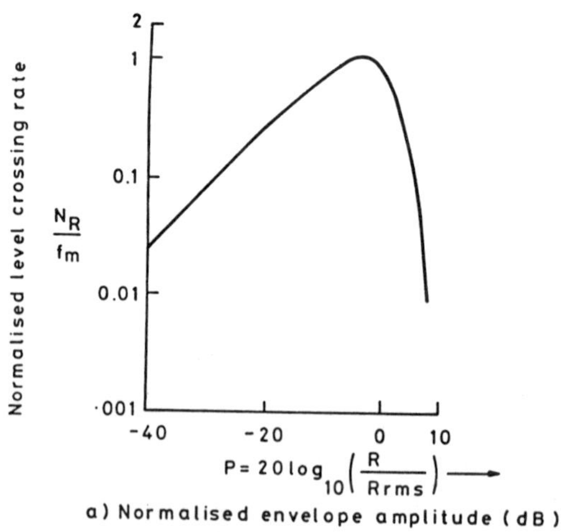

Fig. 3.7(a) Level Crossing Rate

defined as

$$\text{Average duration of fade} = \frac{\text{Probability } r \le R \text{ in time interval} T}{\text{Level crossing rate } N(R)}$$

For a Rayleigh fading envelope, the probability $r \le R$ is

$$P(r \le R) = \int_0^R p(r)dr = 1 - \exp\{-\rho^2\} \tag{3.21}$$

Substituting for N(R) from eqn.(3.20), the average duration of fade for a vertical monopole is given by

$$\bar{\tau}(R) = \frac{\exp\{\rho^2\}-1}{\rho\, f_m \sqrt{2\pi}} \tag{3.22}$$

This expression is plotted in Fig. 3.7b it shows that as the ratio ρ is decreased the average duration of the fades decreases since the signal remains above the level R. However, as the level is increased, a point is reached whence the signal lies below R most of the time and $\bar{\tau}(R)$ can be very long.

3.2.5 Random FM

The time varying phase of the fading signal appears in the form of random frequency modulation. These variations appear as random noise at the output of a FM demodulator. Increasing the transmitter power or receiver sensitivity is

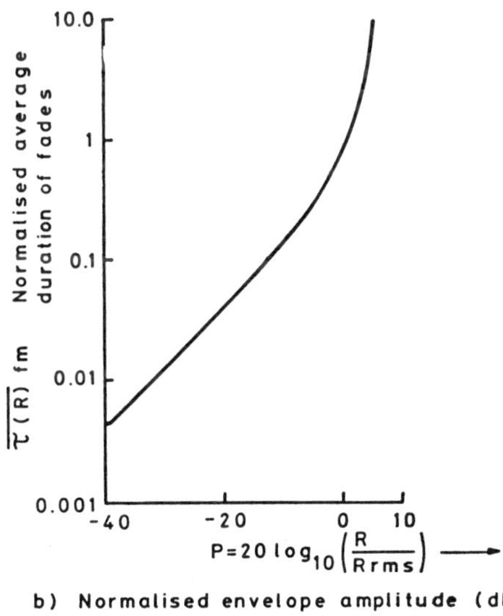

Fig. 3.7(b) Duration of Fades

of little consequence in FM transmissions when random FM dominates the performance. Fortunately random FM is less significant at UHF and below.

Expressions for the p.d.f. for the random FM and the power spectrum are available in Ref(3). In contrast to the sharply defined power spectrum of the signal (Doppler spectrum) the spectrum of the random FM decreases asymptotically as 1/f for f/f_m large and is insignificant beyond $5f_m$ as shown in Fig 3.8. The larger frequency excursions are due to the deep fades. As R is decreased, i.e. for the deeper fades, the frequency deviation will increase proportionately. For a 20dB fade the significant frequency deviations set the random FM bandwidth equal to $10f_m$ and it decreases to f_m when $R = R_{r.m.s.}$

3.3 FREQUENCY SELECTIVE FADING

Fading of the envelope adequately describes the influence of multipath for single-channel voice transmissions. In general the effects of multipath on a 'packet' of frequencies needs to be considered. Consider two frequency components in the message bandwidth. If these components are close in frequency then the two sets of phasors due to the multipath components are nearly alike for the resultant amplitude and phase. Thus these two frequency components will tend to fade in a correlated manner. However, as the frequency separation increases the set of phasors at one frequency tend to decorrelate with the set at the other frequency. This occurs because the relative phases of the phasors are 'jumbled' as the frequency separation becomes large. The extent of decorrelation depends on the time delays and the resultant spread since the phase shifts arise from the excess path-lengths. For large delay spreads the

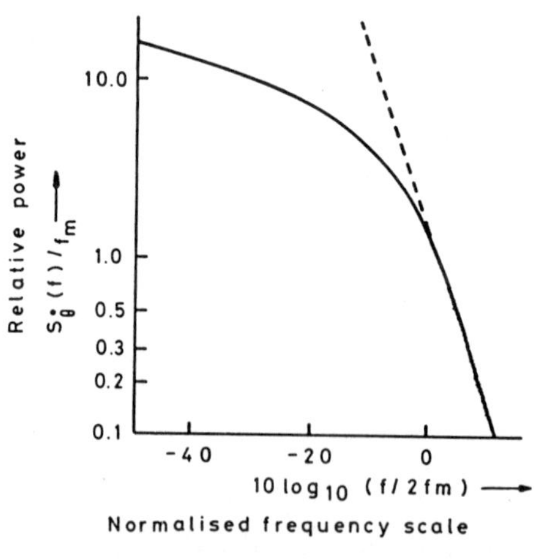

Fig. 3.8 Power Spectrum of Random FM

phases will vary over many radians for smaller frequency separation of the spectral components. Thus signals which occupy a wider bandwidth than this separation will be distorted. In the literature this bandwidth is referred to as the coherence (or correlation) bandwidth since it is defined by the decorrelation of spectral components; the spectrum in fact resembles a fading amplitude as a function of frequency and hence the phenomenon is known as 'frequency-selective fading'.

The wideband characterisation of fading channels has been discussed in (6,8) and experimental data analysed to obtain statistical parameters like the delay spread, average delay and coherence bandwidth (6,7). An illustration of the measured power-delay profile, i.e. echo power versus time delay, and the frequency correlation function are shown in Fig. 3.9. Note that the frequency correlation function is not a smooth function as suggested in the literature [2,3,4].

3.4 LONG TERM FADING

The scattering model describes the "local" statistics of the signal envelope which is completely defined by the "local mean", since the Rayleigh distribution is determined by a

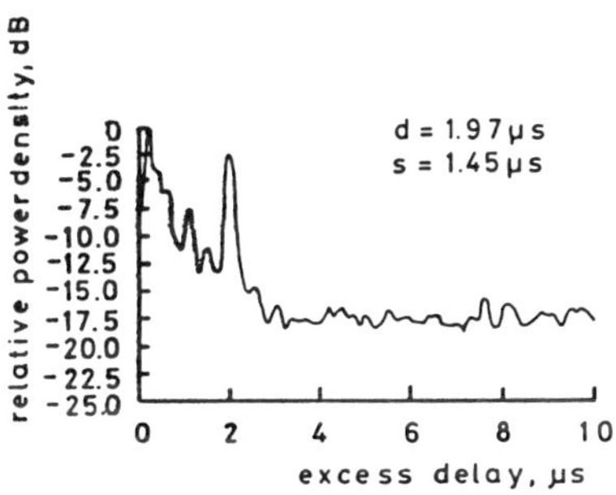

a) Typical average envelope power against time delay for an urban area

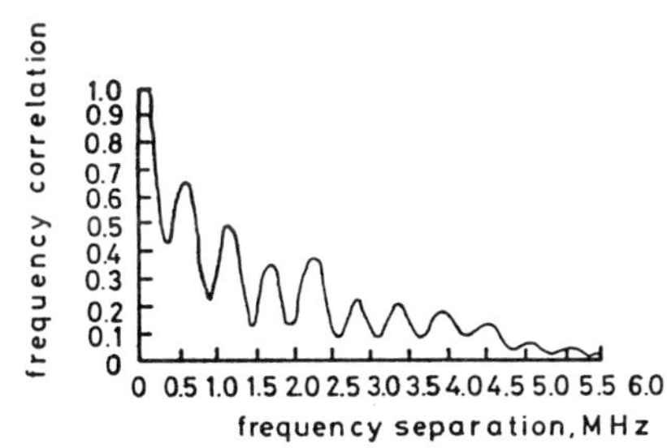

b) Typical frequency correlation function for an urban area

Fig. 3.9 Wideband Statistical Characterising Functions

single parameter "b_o". The local mean is obviously dependent on such factors as the distance from the transmitter, the nature of the terrain, the type of buildings around the mobile, orientation and width of the streets, etc. Hence, the process is non-stationary and may be considered as stationary only over a small distance of vehicle travel where the factors influencing the "local mean" might be judged to have remained sensibly unchanged.

3.4.1 Statistics of the Local Mean

It has been suggested that the distribution of the local mean of the received signal, at fixed range, frequency and within the same environmental class, exhibit an approximately lognormal distribution. Measurements by various researchers support this suggestion. From data collected in London at 168 MHz (9), frequency histogram of the local mean of the received signal is plotted in Fig. 3.10. The cumulative distribution of the same data is plotted on probability paper in Fig. 3.11, the agreement with the log-normal distribution is evident.

The standard deviation σ of the local mean varies between 4 and 8 dB. In general it is higher in open and suburban areas than in urban areas. Within an urban area σ appears to increase with the increase of the proportion of high buildings. The standard deviation is sensitive to the size of the area considered, the larger the area, the higher the value of σ. It has been also noted that σ increases

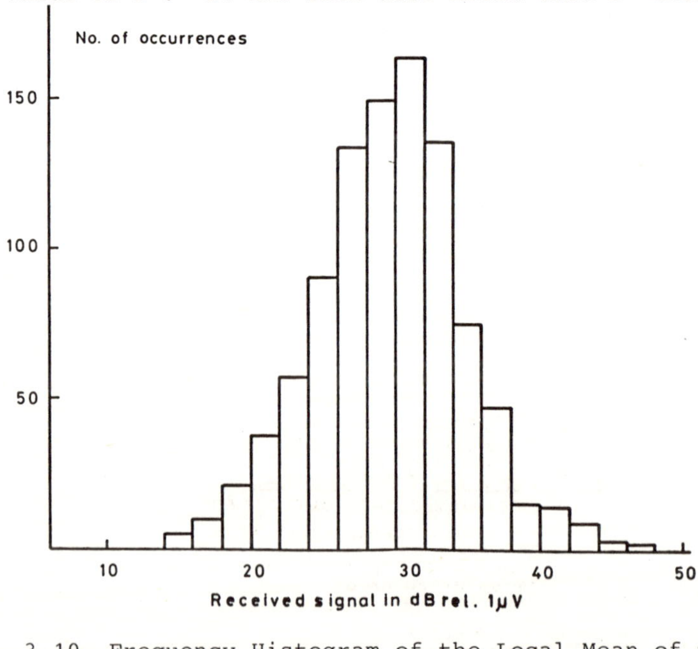

Fig. 3.10 Frequency Histogram of the Local Mean of the Received Signal at 2km Range, 168 MHz

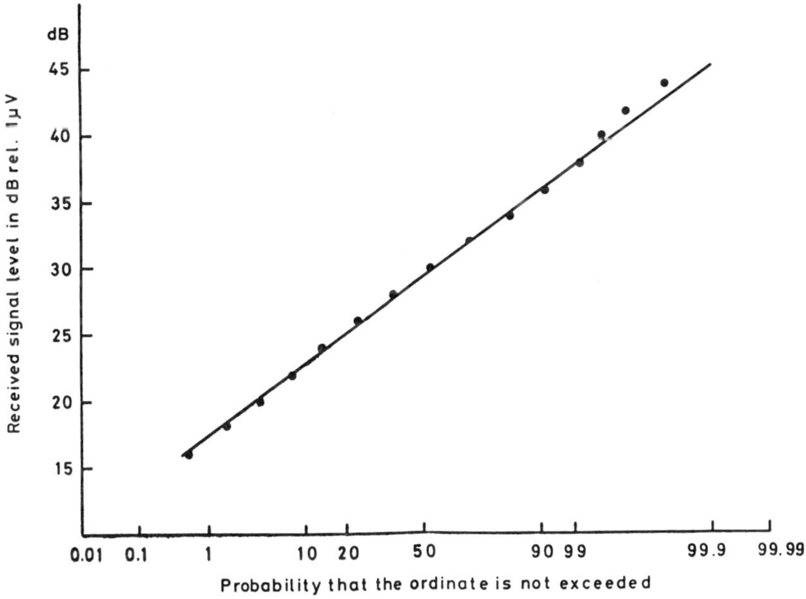

Fig. 3.11 The Cumulative Distribution of the Local Mean of the Received Signal at 2km Range at 168 MHz, Plotted on Probability Paper.

with the increase in transmission frequency.

3.4.2 Factors Influencing the Median Path Loss

The median path loss in urban areas was found to be dependent on a number of factors namely the range, the frequency of transmission, the nature of the urban environment, the base station antenna height and the type of terrain.

3.4.2.1 The variation with range. In general the median received signal is reduced as the mobile moves away from the base station. It has been suggested that the median path loss is approximately proportional to the fourth power of the distance between the base station and the mobile and prediction models based on the theoretical plane earth equation have been proposed. Such models predict the median path loss by adding to the plane earth loss a "clutter factor" which characterizes the particular urban environment (Fig. 3.12). The clutter factor, termed "β", has been found to be frequency dependent.

3.4.2.2 The variation with transmission frequency. Strong correlation between the path loss values at VHF and UHF was noted when measurements were made using the same base

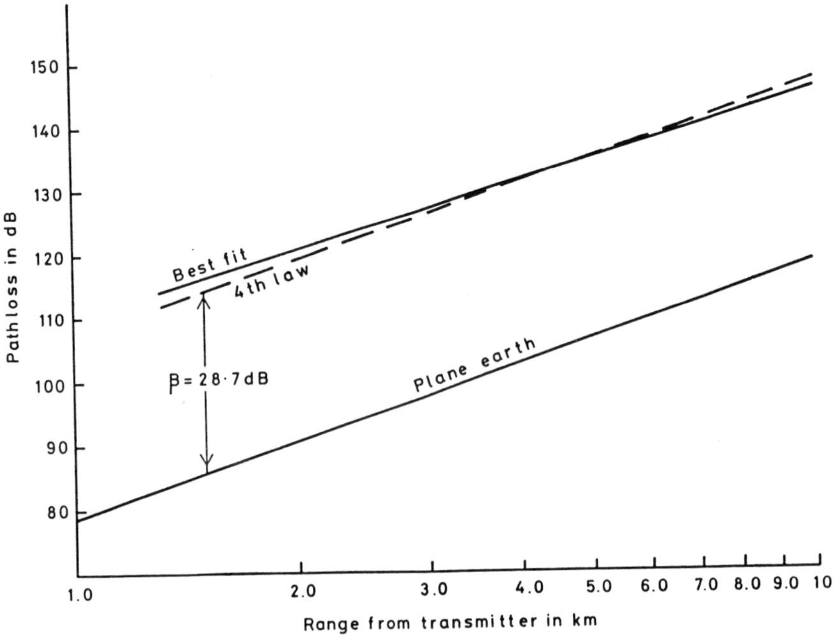

Fig. 3.12 Variation of the Median Path Loss with Range, London, 168 MHz

station. The loss at UHF is consistently higher than that at VHF at all ranges and for all types of environment.

3.4.2.3 The Urban environment. The propagation of radio waves in urban areas is considerably influenced by the nature of urban environment i.e. the size of building, their proximity to each other, the width of roads, the existance of parks, etc. Built-up areas are usually classified into urban and suburban, the urban being loosely defined as dominated by tall buildings, office blocks and high rise residential towers, while a suburb is basically residential where houses with gardens, playfields and parks are the main features. Signals suffer higher attenuation in urban areas than the suburbs, with typical excess loss values of 6 to 10 dB. Some difficulties arise from the qualitative definition of urbanization and some degree of vagueness is unavoidable due to the subjective element associated with such definition. An alternative method has been proposed (9) to define the built-up areas quantitively by means of two parameters, the first is a land usage factor describing the percentage of the area under consideration covered by any buildings. The second parameter is a degree of urbanization factor describing the percentage of high building with a built-up area. Fig. 3.13 displays the relationship between the path loss values, at 168 MHz in London, and the degree of urbanization.

3.4.2.4 Base Station antenna height. A good "launch" is always desirable for a good coverage i.e. the base antenna should clear most, if not all, the obstacles in the immediate vicinity of the base site. While it is intuitive that a higher antenna would provide a wider coverage, the exact relationship between antenna height and the path loss remains unclear. It is acceptable though to employ a relationship extracted from the plane earth model relating the received signal strength in watts to the square of the antenna height. The few measurements reported appear to support such assumption. It is prudent to point out here that the temptation to employ exceedingly high antennas can have adverse effects on channel reuse distances i.e. it can be a self defeating practice as far as spectrum utilization is concerned.

3.5 PREDICTION OF THE MEDIAN PATH LOSS IN BUILT-UP AREAS

Several methods have been proposed for predicting the path loss but only few are intended for application in built-up areas. Some of these are reviewed briefly here.

3.5.1 The Urban Clutter method.

Predictions are based on the formula.

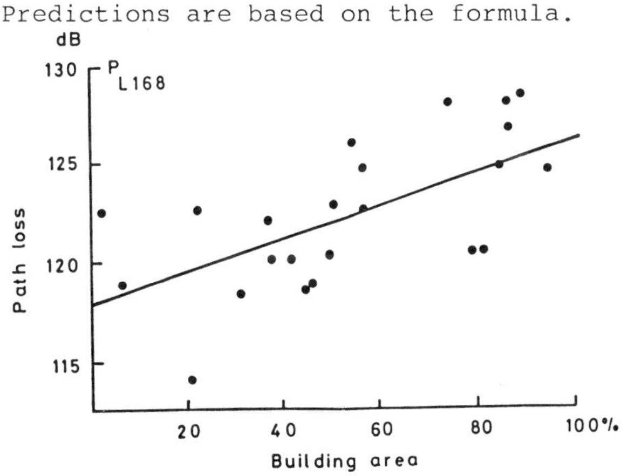

Fig. 3.13 Influence of the Area of Buildings of 4 Floors or More, as a Percentage of the Building Site Area, on the Value of the Path Loss 2 km Data

$$\text{Path loss} = \text{Plane earth loss} + \beta \qquad (3.23)$$

where β is the urban clutter factor, the value of β varies in accordance with frequency, and the nature of the area under consideration. Reported values for β in Birmingham (10) are 19 and 34 dB at 160 and 450 MHz, respectively, while in London values of 29 and 37 dB were measured at the same frequencies (11).

3.5.2 Okumura's Method and Hata's Formulae

Following an extensive series of propagation measurements in and around Tokyo, Okumura (12) produced an elaborate analysis of the results and hence arrived at an empirical method for signal strength prediction. Essentially Okumura's method is based on determining the free-space path loss between the transmitter and the receiver and then adding or subtracting numerous correction factors to account for the nature of the terrain, the extent of urbanisation, the heights of the antenna etc. The basic formulation of Okumura's prediction technique can

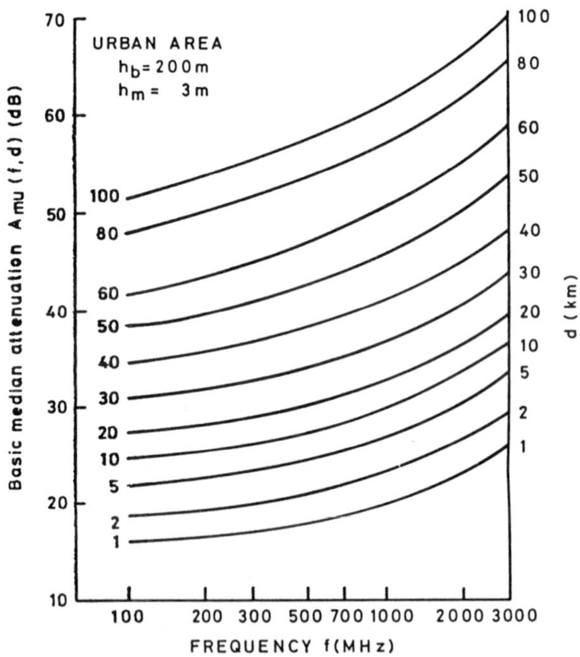

Prediction curve for basic median attenuation relative to free space in urban area over quasi-smooth terrain, referred to $h_b = 200$m, $h_m = 3$m.

Fig. 3.14 Okumura's Method for Path Loss Calculation

be expressed as

$$\text{Path loss} = L_F + A_{mn} - H_{tn} - H_{rn} \text{ dB} \tag{3.24}$$

In this expression L_F is the free-space path loss and A_{mn} is the median attenuation relative to L_F in urban areas over what is defined as "quasi-smooth" terrain with a transmitter antenna height h_t of 200m and a receiver antenna height h_r of 3m. A_{mn} is a function of frequency and range and is expressed in graphical form by the series of curves shown in (Fig. 3.14), H_{tn} and H_{rn} are correction factors to account for aerials not at the reference heights of 200m and 3m; they are called the height-gain factor and are also functions of frequency and range. Okumura's paper contains graphs from which the appropriate values for any specific situation can be extracted.

If the terrain cannot be treated as "quasi-smooth" or if the environment is not urban, then further corrections have to be made. In fact Okumura produced eight factors intended to correct for suburban and open areas, sloping terrain, hilly terrain, mixed land sea paths etc., all these factors being expressed in graphical form. Street orientation in urban areas can also be taken into account.

The method is often quoted since it is almost the only technique readily available that is specifically intended for use in built-up areas and which takes account of variations in both terrain and urbanisation. However, it is rather involved and difficult to implement directly on a computer. This latter difficulty arises because since the data is available only in graphical form, this has to be entered point by point into a computer memory, and interpolation routines have to be devised for obtaining intermediate points. A further reservation about the method stems from the differences, topographical and otherwise, between Tokyo, where the measurements were taken, and European cities.

Nevertheless, in order to put Okumura's technique into a form suitable for implementation on a computer, Hata (13) recently published an empirical formula for the propagation loss which is based on Okumura's paper. The path loss expression so produced is

$$L_{path} = 69.55+26.16\log f-13.82\log h_t-A(h_r)+(44.9-6.55\log h_t) .\log d \tag{3.25}$$

where f lies in the range 150- 1500MHz

h_t lies in the range 30- 300m

and d lies in the range 1-20 km.

$A(h_r)$ is the correction factor for the mobile antenna height and is computed as follows:

For a small or medium city:

$$A(h_r) = (1.1 \log f - 0.7)h_r - (1.56 \log f - 0.8) \text{ dB} \quad (3.26)$$

where h_r lies in the range 1-10m.

For a large city

$$A(h_r) = 8.29(\log(1.54\, h_r))^2 - 1.1 \text{ dB}; f \leq 200 \text{ MHz}$$

and (3.27)

$$A(h_r) = 3.2(\log(11.75\, h_r))^2 - 4.97 \text{ dB}; f \geq 400 \text{ MHz}$$

3.5.3 Ibrahim and Parsons Method

The method (14) is designed to predict the median path loss in flat urban areas. The predictions are made for 0.5 km squares; ordnance survey maps are required to extract the average terrain height for each square and its range from the transmitter. The methods also require land usage data obtainable from the local Authorities. Two models were developed, the first (empirical) gives the median path loss in dB between isotropic antennas as:

$$P_L = 47.7 - 8 \log h_r - 20 \log f\, h_t + \left(40 + 14.15 \log\left(\frac{f+100}{156}\right)\right) \log d$$

$$+ 0.265L - 0.37H + K \quad (3.28)$$

where $K = 0.087U - 5.5$ for the highly urbanized centre of the city otherwise $K = 0$.

h_r, h_t = Mobile and base antenna heights in metres respectively

f = Frequency in MHz

d = Range from base station in metres

L = Land usage factor

U = degree of Urbanization factor

H = difference in spot Height between the base and the mobile.

The second model, (semi-empirical), gives the median path loss as

$$P_L = \text{Theoretical Plane earth loss} + \beta \quad (3.29)$$

where $\beta = 20 + \frac{f}{40} + 0.18L - 0.34H + K$

when $K = 0.094\, U - 5.9$ for the highly urbanized city centre, otherwise K = 0.

Prediction methods in general compute the median path loss and it is up to the system designer, having set his "quality" of coverage criterion, to employ the statistics of the signal envelope in order to meet his requirement.

REFERENCES

1. Ossana, J.F., Nov. 1964, 'A Model for Mobile Radio Fading due to Building Reflections: Theoretical and Experimental Fading Waveform Power Spectra", BSTJ, 43, pp 2935-2971.

2. Clarke, R.H., July-August 1968, 'A Statistical Theory of Mobile Radio Reception', BSTJ, 47, pp 957-1000.

3. Jakes, W.C., 1974, 'Microwave Mobile Communications', Wiley Interscience.

4. Gans, M.J., Feb. 1972, 'A power Spectral Theory of Propagation in the Mobile Radio Environment', IEEE Trans. V.T., 21, pp 27-37.

5. Rice, S.O., Jan. 1948, 'Statistical Properties of a Sine Wave plus Random Noise', BSTJ, 27, pp 109-157.

6. Bajwa, A.S., 1979, 'Wideband characterisation of UHF Mobile Radio Propagation in Urban and Suburban Areas', Ph.D. thesis, Department of Electronic and Electrical Engineering, University of Birmingham.

7. Bajwa, A.S., April 1982, 'Small-area Characterisation of UHF Urban and Suburban Mobile Radio Propagation', Proc. IEE, 129, pp. 102-109.

8. Parsons, J.D. and Bajwa, A.S., April 1982, 'Wideband Characterisation of Fading Mobile Radio Channels', Proc. IEE, 129, Pt. F, No. 2, pp. 102-109.

9. Ibrahim, M.F.A., 1981, 'Signal Strength Prediction for Mobile Radio Communication in Built-up Areas', Ph.D. Thesis, University of Birmingham.

10. Allsebrook, K., 1977, 'An investigation of the propagation of Radio waves at Frequencies in the VHF and UHF bands within certain British Cities', Ph.D. Thesis, University of Birmingham.

11. Parsons, J.D., Ibrahim, M.F.A. and Samuel, R.J., Feb. 1980, 'Median Signal Strength Prediction for Mobile Radio Propagation in London', Elec. Letters, 16, No. 5, p. 172.

12. Okumura, Y. et al, Sept. 1968, 'Field Strength and its Variability in VHF and UHF Land Mobile Service', Rev. Elec. Comm. Lab., 16, pp 825-873.

13. Hata, M., August 1980, 'Empirical Formula for Propagation Loss in Land Mobile Radio Services', IEEE Trans. V.T., 29, pp 317-325.

14. Ibrahim, M.F.A. and Parsons, J.D., August 1983, 'Signal Strength Prediction in Built-up Areas', Part I, Proc. IEE, Part F, 130, No. 5, pp 377-385.

Chapter 4

Aerials and base station design

J.G. Gardiner and R.E. Fudge

4.1 INTRODUCTION

The importance of minimising spectrum pollution in the congested radio environment generally is self evident and as demands on spectrum allocations for mobile radio continue to grow the requirements of Home Office specifications for transmitter and receiver performance respond by defining progressively smaller unwanted emission characteristics from equipments in isolation. However, in the practical system situation the needs of many users to independently achieve good radio coverage of the same geographical areas dictates that many systems must share common facilities for transmission and reception at favourably located rooftop or hill top sites and this need greatly extends the demands made on base station engineering if the performance of transceivers in isolation is to be preserved in co-site operation. The material which follows considers the co-siting problem by reviewing the characteristics of site system components, highlighting aspects of the characteristics of these components which dictate complete site system design and considering the general problem of predicting the location in frequency and magnitude of unwanted emissions from communal sites which would be absent from the spectrum if the characteristics of the transceivers concerned in isolation could be ideally maintained.

Essentially, whilst the mechanisms by which interactions among co-sited transmitters are numerous and often complex, they fall into two basic categories.

1) Those which derive from coupling effects between the equipments installed at the site via radiated interaction among aerials, feeders, mains power supplies and so on which result in nonlinear transmitter output stages being required to handle signals in addition to the carrier they were designed to transmit, either in a transmission mode or a reflection mode.
2) Those which are associated with the metal mast structure which supports the antennas and other metal work on or around the site which develop corroded joints and contacts whose characteristics are non-linear and which generate and re-radiate unwanted spurious when illuminated by the transmit antennas.

The former category of mechanism can reasonably be regarded as being under the direct control of the site engineer and it is his responsibility to ensure that adequate standards of spectral purity are achieved, while arguably the latter may be outside his control. He will also be subject to cost constraints and will attempt to attain a minimum acceptable standard of performance with minimum expenditure on equipment so that an appreciation of how spurious generation arises and how it may be controlled is of considerable importance.

4.2 AERIALS AND FEEDERS

4.2.1 Antennas

The simplest approach to co-siting is to employ a separate single antenna for each base station transmitter and receiver on the site. (This is often convenient from the organisational point of view if no individual user has responsibility for the communal facility) and the characteristics of typical antennas and feeders are therefore an obvious starting point. Generally there are only two common requirements from base station antennas - to obtain omni-directional or wide-angle sector cover for access to mobiles and highly directional operation for fixed link work. Since mobile radio frequency allocations are confined to relatively narrow bands, broad band antennas are seldom required. Whip antennas are universally used on vehicles and this defines vertically polarised emissions for the omni-directional antennas and a typical communal site will be seen to carry a selection of simple folded dipoles for communication with vehicles. For link applications yagi antennas are usually employed to obtain the necessary directivity and may be vertically or horizontally polarised.

The theory of antenna design is beyond the scope of the present work and the reader is referred for details to any of several comprehensive texts available (1-3), for completeness however the characteristics of some common antennas are reproduced in Fig. 4.1. (It should be noted that antenna gain is sometimes quoted in dB relative to a half wave dipole (dBd) or relative to a hypothetical isotropic radiator (dBi) the difference being 2.2 dB).

Variations on the elementary folded dipole are available to achieve either antenna gain or various degrees of directivity or both, the commonest techniques being "stacking" and "baying" of dipoles by which several radiators are connected to a single transmitter with carefully designed phase shifts between them (achieved by specified lengths of additional feeder cable). An extension of the stacking principle results in the co-linear structure with omnidirectional gain (typically 3-9 dB), baying may also be used to produce a null in the polar diagram whose position may be chosen to reduce interference from a particular source.

It is evident from the quoted performance figures that limitations to gain and directivity exist which are related to physical dimensions so that gains for stacking etc. are limited generally to around 6 dB at VHF and 10 dB at UHF for

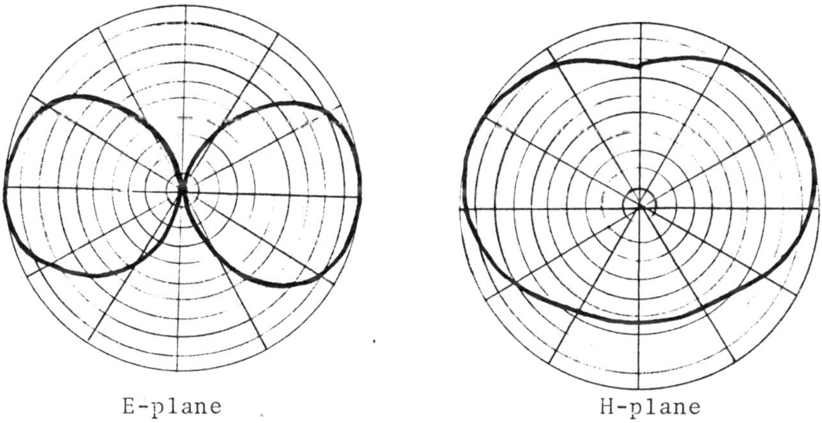

(a) Folded dipole (assymetry in H-plane caused by mast) bandwidth ± 12%

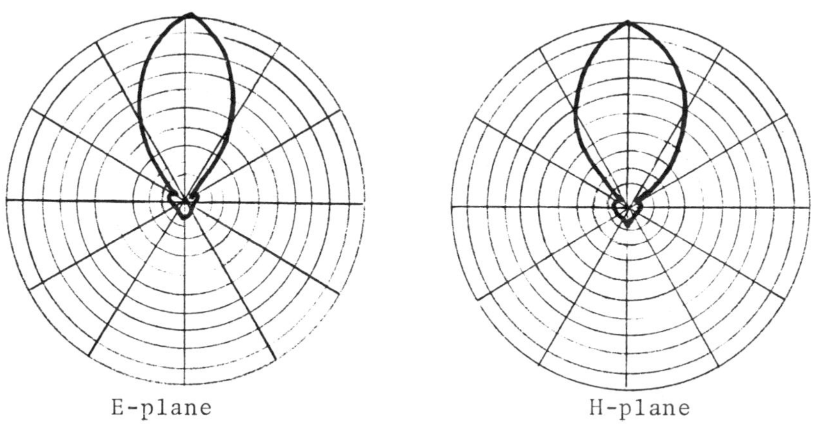

(b) 8-element yagi , bandwidth ± 4%

Fig. 4.1 Typical polar diagrams for folded dipole and yagi antennas

the omnidirectional case, a practical limit at UHF of 18 elements for yagi structure gives a gain of 14 dBd approximately (see Table 4.1)

4.2.2 Feeders

Feeders are often taken for granted in site engineering but the range of performance, and cost, of available cable is wide and, to some extent, feeder performance can significantly influence the basic choice of site configuration. Obviously, low loss is a desirable feature but an equally desirable aspect, that of screening effectiveness or the cable's capacity to keep wanted signals inside the outer sheath, is sometimes neglected. In the primitive site with a separate antenna for every transmitter and receiver a common 50Ω feeder choice is Uniradio 67 and this may be regarded as a minimum standard for a satisfactory installation at frequencies up to 500 MHz. Whilst this is a low cost cable its performance is by no means good - loss at 100 MHz is typically 6.8 dB per 100m rising to 15 dB or so at 500 MHz. This loss is clearly a substantial drawback in the transmit path but can often be compensated for by antenna gain but it is even more serious in the receive path since the feeder loss adds directly to effective receiver noise figure. Moreover, braided cables radiate (loosely braided coaxial cable can be used quite effectively for 'leaky feeder' in some situations) so that if several braided feeders are run together up a mast, a measure of coupling among them can be expected. Better results are achievable with cables with solid outer conductors and generally as diameter increases loss decreases. Table 4.2 compares the loss characteristics of a number of solid outer cables with, in one case, a solid ptfe dielectric and in the other predominantly air dielectric with the centre conductor supported by a high density dielectric helix. Naturally, higher performance means higher cost and the 7/8" diameter solid dielectric feeder currently markets at around 20 times the cost of UR 67.

4.3 ISOLATING COMPONENTS

As the number of users at a communal site increases the problem of operating separate antennas for each transmitter and receiver becomes more acute. In particular physical

TABLE 4.1 Gains and beamwidths of yagi antennas

No of elements		2	3	4	6	8	12	18
Gain (relative to $\lambda/2$ dipole,) dB		3	6	7.5	8.5	10	12	14.6
Beamwidth half power	E plane °	62	60	58	56	40	30	23
	H plane °	95	80	74	64	50	36	32

TABLE 4.2

Cable outer diameter (inches)	Loss per 100 m (dB)		
	100 MHz	500 MHz	
$\frac{1}{4}$	4.5	10.4	solid dielectric
$\frac{3}{8}$	3.5	8.3	
$\frac{1}{2}$	2.4	5.8	
$\frac{7}{8}$	1.36	3.5	

Cable outer diameter (inches)	Loss per 100 m (dB)		
	100 MHz	500 MHz	
$\frac{3}{8}$	2.8	6.6	helix dielectric
$\frac{5}{8}$	1.75	4.1	
$\frac{7}{8}$	1.18	2.7	
$1\frac{1}{8}$	0.91	2.1	
$1\frac{5}{8}$	0.64	1.47	

separation between antennas decreases so that (a) receivers experience desensitisation resulting from high level inputs from transmitters, (b) transmitters begin to couple substantial amounts of energy into each others output ports via coupling between antennas, feeders, etc. resulting in intermodulation products being transmitted together with wanted carriers, (c) the total transmitted power level incident on mast metal work goes up and rusty bolt effects become important as sources of spectrum pollution. Moreover, if the cost of high quality low loss feeders becomes very high, then it is plainly uneconomic to instal many runs from equipment room to mast head so that some measure of combining or at least supplementing isolation from antenna spacing must be employed. The possible system components are filters (cavity or helical resonator types), ferrite isolators and circulators, hybrid combiners, duplexers, receiver distribution systems and Butler networks.

4.3.1 Filters

The commonest form in use is the quarter wavelength coaxial cavity operating in the TEM mode, see Fig. 2(a). This is very satisfactory at high-band VHF frequencies but becomes rather bulky at lower frequencies. At UHF frequencies the coaxial cavity is usually used, although for an unloaded Q_o of greater than 15000 a hollow metallic cubic resonator has been developed in Japan.

Simple design data have been published (4) for the quarter-wave cavities, which shows that the diameter in inches of the cavity (D), the Q_o and the resonant frequency f_o are related by

$$Q_o \simeq 74D\sqrt{f_o}$$

and gives a graph of the loaded Q_L (effective Q in circuit) as a function of the unloaded Q_o and the through loss. The quarter-wave cavity has the advantage of good temperature stability if the length of the centre conductor is controlled by an invar rod.

At low VHF frequencies electrically-short coaxial cavities become attractive as do cavities using a slow-wave resonating structure, both designs being more compact than quarter-wave resonators. One problem concerning some designs of electrically-short cavities is that of ensuring adequate temperature stability over the working frequency band. Tests with multicylinder cavities have however been encouraging, good stability having been maintained over a tuning range of ± 10% of the centre frequency. Generally, the quarter-wave and the electrically short coaxial cavities cater for a range of Q_o from 1000 to 5000, more usually over the range 3000-7000.

For an unloaded Q_o in the range of 1000 to 3000 it is usual to employ cavities with helical resonators - see Fig. 2(b), and these may also be used as bandstop resonators to remove a particular unwanted frequency in an integrated site system. Within this range, helical resonators offer a higher ratio of unloaded Q_o to volume than comparable coaxial resonators.

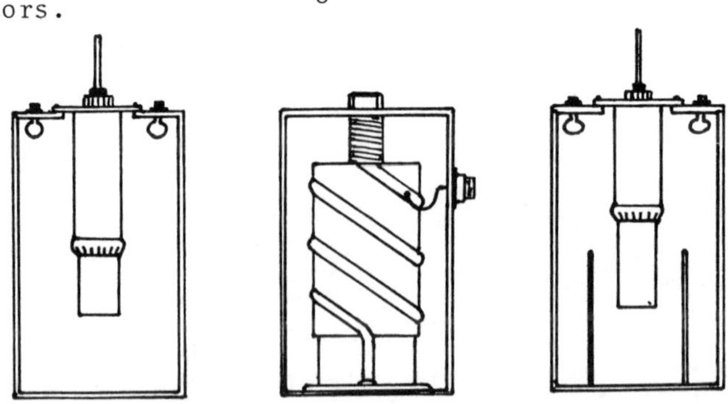

(a) (b) (c)

Fig. 4.2 Resonator structures : a) λ/4 coaxial, b) helical, c) multicylinder

Helical resonators may be operated as half-wave transmission lines, as shown in Fig. 2(b), or as quarter-wave lines, where one end of the resonator is open-circuited.

Cavity resonators have to be provided with some form of tuning to allow them to operate with a range of transmitter frequencies. With the coaxial structures, this is most easily accomplished by making the centre conductor telescopic. Since the resistive loss in the cavity is largely concentrated on the centre conductor, silver plating is usually employed on the two tubes, which are electrically connected by sliding silver-plated spring fingers. The helical resonators may be tuned by a lumped capacitance at the end of the quarter-wave line, or by metallic plugs screwed in or out through the cavity wall.

4.3.2 Ferrite Isolators

These devices are readily available for VHF and UHF bands and will provide 20 dB of isolation and about 1 dB insertion loss with the aid of suitable external resonating capacitor networks. The bandwidth over which isolation is maintained is generally narrow (about 5%) and care is required in using tandem connection of ferrite devices if total isolation approaching the sum of individual contributions is to be achieved.

4.3.3 Hybrid Combiners

These are essential combining components for transmitters which are very closely spaced in frequency (i.e. closer than 100 KHz). Various configurations are available (5) usually realised in coaxial line and isolations of the order of 30 dB can be readily achieved over rather limited bandwidth, comparable to the isolators above. The disadvantage of hybrids as transmitter combiner devices is the inherent 3 dB minimum insertion loss associated with every combination of two inputs, i.e. 10 dB total insertion loss in any input/output path for an 8-way hybrid combiner.

4.3.4 Duplexers

These fall into two principle categories, 'stop-band' and 'pass-band', designations which indicate the filter configurations associated with both transmit and receive paths. Stop-band duplexers insert a band stop filter at the transmit frequency in the receiver input path to prevent transmit energy from blocking the receiver and a similar filter at receive frequency in the transmit path to prevent received signal power from dissipating in the transmitter output port. Band pass duplexers achieve the same results by introducing band pass filters at appropriate frequencies transmit and receive paths. A typical duplexer requirement would set a target insertion loss in both transmit and receive paths of around 1 dB while ensuring that 60 dB rejection of transmit power is achieved in the receive path. This target can generally be met with helical or cavity resonator filters and is somewhat easier to achieve with the band stop configuration (6). Nevertheless, the band pass duplexer plays an important role in many cases since this device attenuates transmitter noise sidebands in the receiver band where the stop band types do not.

4.3.5 Receiver Distribution Systems

The receiver distribution system comprises bandpass filters, a low-noise single-stage amplifier and a hybrid splitter to provide the requisite number of outlets.

The overall noise figure of the receiving system can be expressed as

$$F_o = F_2\ell_1 + \frac{(F_3-1)}{G_2}\ell_1 + \frac{(F_{RX}-1)}{G_2}\ell_1\ell_3$$

where

ℓ_1 is the insertion loss of the bandpass filters,
F_2 is the noise figure of the amplifier,
G_2 is the gain of the amplifier,
F_3, ℓ_3 are the noise figure and insertion loss of the hybrid-splitter

and F_{RX} is the noise figure of the base station receiver.

It would be highly desirable if an improvement in signal reception could be achieved from a receiver distribution system and it therefore follows that $F_o < F_{RX}$ is required. Returning to the expression for F_o, the optimum values of the terms F_2, G_2 and ℓ_3 can be stated with certainty. For example, the insertion loss, ℓ_3, of an 8-way splitter is approximately 9.5 dB, the gain, G_2, of a single-stage amplifier at VHF and UHF is in the region of 17 dB and its noise figure, F_2, for a conjugate match at 50Ω is typically 3.5 dB. This leaves the variable ℓ_1 to be determined for a given receiver noise figure, F_{RX}.

Using the values stated above and the condition $F_o < F_{RX}$ becomes

$$\ell_1 < \frac{F_{RX}}{(2.22 + 0.176\,F_{RX})}$$

and for F_{RX} = 10 dB which is typical for the majority of VHF/UHF receivers, it follows that the insertion loss ℓ_1 of the input filter must be equal to or less than 4 dB.

4.3.6 Butler Networks

Recalling the comments above relating to hybrid combiners it is apparent that the loss associated with each two input combination does not derive from the network itself but from the way in which the hybrid divides input power equally between the output port and a fourth port which is terminated in the hybrid design impedance (usually 50Ω). An ideal system would retain this power and use it to feed a further antenna and this arrangement may be employed with combinations of four transmitters to produce a Butler matrix, fig. 3. If quarter-wave hybrids are used then the outputs of the Butler matrix have quadrature phase relationships but by spacing directional antennas appropriately around a mast a far-field polar diagram approaching the omni-directional can be achieved (7).

4.4 INTERMODULATION PRODUCTS - HOW MANY AND WHERE

The foregoing notes have summarised the available hardware from which a site system can be assembled. It is to be expected, however, that residual transmitter coupling and rusty bolt effects will result in the generation of some unwanted spectrum components. This section treats the problem of identifying how many significant products may be anticipated and what their frequencies will be.

4.4.1 General Representation of a Nonlinearity

One method of representing a nonlinear device is by means of a polynomial transfer characteristic

$$y = A_o + A_1 x + A_2 x^2 + A_3 x^3 + \ldots\ldots\ldots A_n x^n \qquad (4.1)$$

where the exciting or input quantity is x, the response or output is y and $A_o A_1$ etc. are constant coefficients. For a truly linear system all coefficients except A_1 are zero, and for practical systems which aim to be linear it is usual to find $A_o = 0$, $A_1 > A_2 > A_3 > A_4$ etc. For the systems under

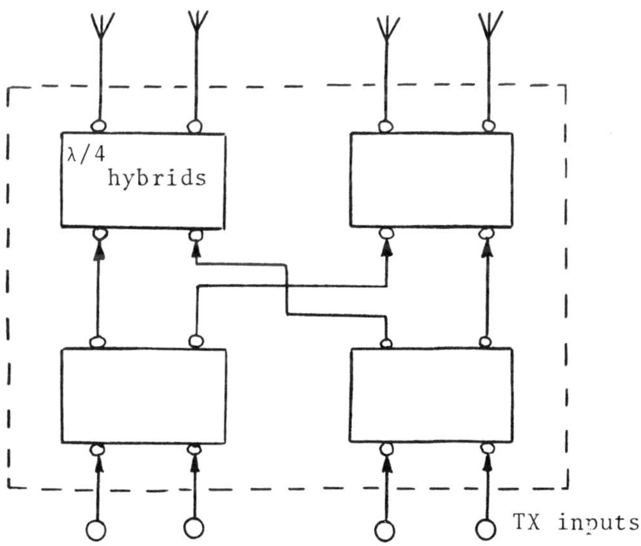

Fig. 4.3 4-way 'Butler matrix' combiner

consideration here there is evidence that this does not always hold and one or more higher order terms can exceed the magnitude of lower order ones.

Equation 4.1 can be written more generally as:-

$$y = \sum_{n=o}^{d} y_n = \sum_{n=o}^{d} A_n x^n \qquad (4.2)$$

where d is the degree of the polynomial (the highest value of n it is considered necessary to include) and n is the order of the constituent component.

4.4.2 Two Carrier Excitation

If the excitation x is composed of two components such that $x = r + s$, then the nth order response is

$$y_n = A_n (r+s)^n \qquad (4.3)$$

The form of the expansion of this for the first four orders is given in Table 3. The coefficients for this table are most easily derived from Pascal's triangle.

If the two exciters are of sinusoidal form then the polynomial representation of a nonlinearity can be used to predict the intermodulation frequencies. To this end we let

$$r = a \cos (\omega_1 t + \alpha), \qquad s = b \cos (\omega_2 t + \beta)$$

these can be inserted in the expanded form of y_n and the amplitudes and locations of the intermodulation products derived. Neglecting for the moment the problem of amplitude prediction, to just predict locations the A_n, a, b, α, β terms can be ignored, this then leaves the expansion in the form

$$Y_n = \sum_{q=1}^{n} \cos (q-n)(\omega_1 t) . \cos q(\omega_2 t)$$

where Y_n represents the frequency components of y_n this can be simply, but tediously, expanded by the use of trigonometrical relationships. If this is performed, and at the same time amplitudes are again discarded together with a change to non-angular frequencies

($f_z = \frac{\omega_z}{2\pi}$), then the relationship can be stated succinctly as

$$F_n = \sum_{m=1}^{n} g_m f_m$$

where F_n is understood to mean the intermodulation frequencies of order n, g_m has the value ± 1 and f_m is either f_1 or f_2. Thus F_n consists of all possible combinations of $\pm f_1$, $\pm f_2$ taken n at a time.

By way of example consider $n = 2$ (the square-law mixer case) then, taking f_2 as being higher than f_1 and listing only the positive frequency results gives

TABLE 4.3

Order n	Expansion of $\frac{Y_n}{A_n}$
1	$r + s$
2	$r^2 + 2rs + s^2$
3	$r^3 + 3r^2s + 3rs^2 + s^3$
4	$r^4 + 4r^3s + 6r^2s^2 + 4rs^3 + s^4$
5	$r^5 + 5r^4s + 10r^3s^2 + 10r^2s^3 + 5rs^4 + s^5$

$$F_n = (f_1+f_1) + (f_1-f_1) + (f_1+f_2) + (f_2-f_1) + (f_2+f_2) + (f_2-f_2)$$

$$= 0 + (f_2-f_1) + 2f_1 + (f_2+f_1) + 2f_2$$

also for n = 5

$$F = 5f_1 + 5f_2 + 3f_1 + 3f_2 + f_1 + f_2$$

$$+ (4f_1-f_2) + (4f_1+f_2) + (4f_2-f_1) + (4f_2+f_1)$$

$$+ (3f_1-2f_2) + (3f_1+2f_2) + (3f_2-2f_1) + (3f_2+2f_1)$$

$$+ (2f_1-f_2) + (2f_1+f_2) + (2f_2-f_1) + (2f_2+f_1)$$

$$+ (2f_2-f_1) + (2f_2+f_1) + (2f_1-f_2) + (2f_1+f_2)$$

It should be noted from this that the coefficients do not always add to n, they may sum to n, n-2, n-4 etc. This arises when a frequency has associated with it both positive and negative signs deriving from different values of g_m. Thus, $f_1 + f_1 - f_1 - f_2 - f_2$ gives $2f_2-f_1$.

It should be noted therefore that the intermod frequencies of order n include all those of order n-2. It follows that an intermod found at frequency F which can be calculated as an nth order may in fact be a n+2Nth product (N is an integer).

4.4.3 Multiple Carrier Excitations

When there are more than two exciters then the treatment can follow that given above with the outcome that the nth order products are derived by taking all possible combinations of the exciting frequencies n at a time:

Some examples for the 5th order case are therefore:

$$f_1 + f_2 + f_3 + f_4 + f_5$$

$$f_1 + f_2 + f_4 - 2f_3$$

$$3f_5 - (f_2 + f_3)$$

The spectral range covered by the intermod extends from, at its lower end, zero frequency in the case of even orders or the smallest difference between two carriers in the odd order case, to n times the highest frequency at the upper end.

4.4.4 Carrier Modulation

A modulated carrier can be considered as a cluster of exciting sinusoids covering the range f-Δ to f +Δ (Δ is the half bandwidth required by the modulation, for AM it is the highest modulating frequency. Each component can be considered as one of the input quantities. The effect of this is to broaden the intermod spectral line of the unmodulated carrier by an amount which is dependent on the bandwidth occupied by the modulations of the contributing carriers and the multiplication factor of that contribution. An example can be shown for one of the cases considered in the previous section of a 5th order product: $3f_5 - (f_2+f_3)$. This will be broadened by $\pm (3\Delta_5+\Delta_2+\Delta_3)$ where Δ_j is the bandwidth occupied by the modulated carrier of frequency f_j. If all the modulations occupy the same bandwidth then the maximum spread of an nth order product will be $\pm n\Delta$ about the product frequency.

4.4.5 Narrow Band Systems

The narrow band of the heading is not in the sense of modulation but the range of carrier frequencies likely to be encountered. Fig. 4.4 shows the spectral distribution of intermod products for orders 2 and 3 for the case of two unmodulated carriers f_1 and f_2. The amplitudes are shown diagramatically as decreasing with order number as an aid to clarity rather than indicating their actual magnitude, and the order number of the product is also shown. It will be seen that the order 2 products are thrown well clear of the exciting frequency whereas the order 3 ones can be reflected back to lie close to the originators. This behaviour difference can be generalised to show that, where the exciters are close together in frequency (i.e. narrow band), then only the odd order products will produce intermods close to them (in band).

These in-band products are of particular interest in most systems and particularly so in the case of land mobile radio where base station transmitter products can affect either the mobile receiver or the return path base receiver.

4.4.4 Mobile Receiver Influence

The front end of the mobile receiver is to a degree non-linear and this nonlinearity will generate intermodulation products. Such a product, generated by receiving in the mobile two or more base station carriers, can fall on a frequency assigned for reception by that mobile, and if strong

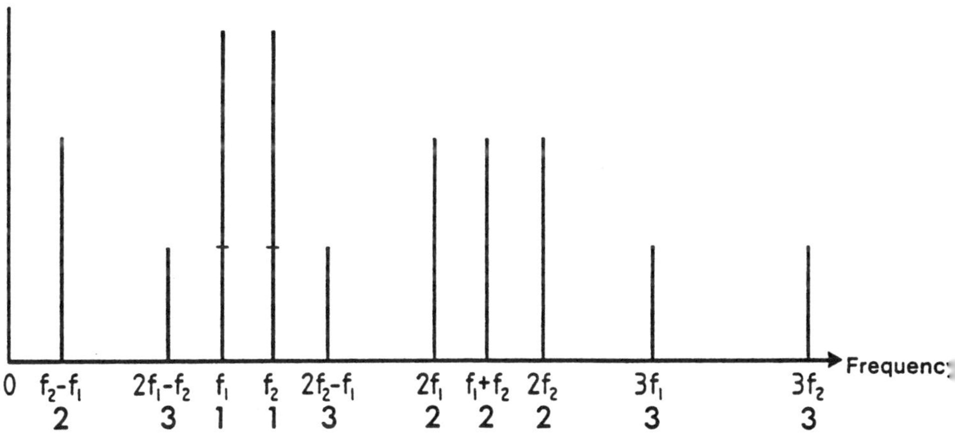
Frequency
0
f_2-f_1
2
$2f_1-f_2$
3
f_1
1
f_2
1
$2f_2-f_1$
3
$2f_1$
2
f_1+f_2
2
$2f_2$
2
$3f_1$
3
$3f_2$
3

Fig. 4.4

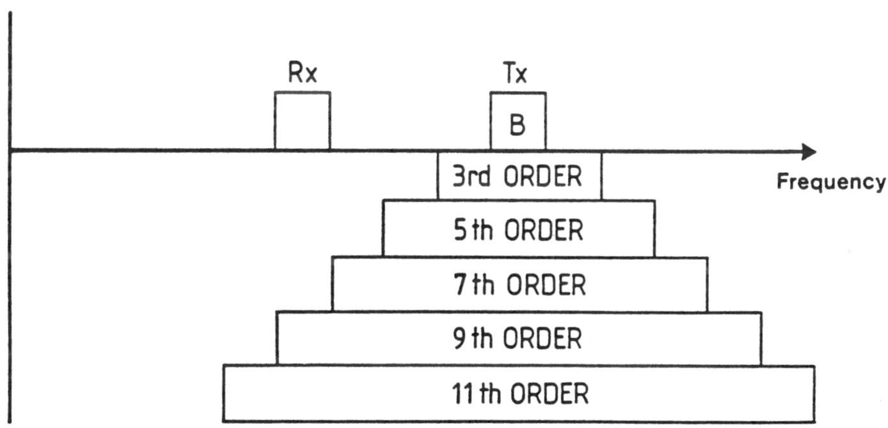
Rx
Tx
B
Frequency
3rd ORDER
5th ORDER
7th ORDER
9th ORDER
11th ORDER

Fig. 4.5

enough can open the mute in the absence of the desired signal, causing severe annoyance to the user. For any given degree of receiver nonlinearity the carriers will be strong enough to do this if the mobile is close enough to the base station (remember the working dynamic range of the mobile i of the order of 100 dB which is greater than the carrier to intermod level for high level carriers) thus there will be a region round the base station in which the mobile could not operate if it decided to avoid this effect. Since the region could well be part, and possibly an important part, of its desired coverage it is usual to choose base station frequency assignments such that no two or three channels form a 3rd order product on another channel, or even so that no 2, 3, 4 or 5 channels form a 5th order product on another channel.

Tables or suitable spacings have been produced for up to 9 channel 3rd order by Edwards, Durkin and Green (8). These are closest spacing (i.e. minimum bandspread) solutions, subsequent (as yet unpublished) work as part of a frequency allocation exercise by PE Consultants for the Home Office has shown that it is not difficult to generate up to 13 channel 3rd order free sequences but these are not necessarily minimum bandwidth solutions. This work shows that to allocate 13 channels free of 3rd order intermods requires a frequency span equal to 127 channels. If a frequency span greater than this is available then more sequences are possible, for instance 66 sequences of 13 channels were found in a spread of 150 channels. Table 4.4 shows channel allocations for 3-8 and 13 channel sequences.

4.4.7 Base Station Receiver Assignments

The base station receiver is also vulnerable to intermods. Those which are generated in the transmitter equipment, and those which would be generated in the receiver, are in principle amenable to treatment with suitably located and designed filters. Neither transmitter nor receiver filtering can affect however the intermods generated by site nonlinearities. The treatments here are either to quench the intermod at source (a subject in its own right) or to take account of them when allocating transmitter and receiver bands of operation.

TABLE 4.4 3rd Order Free Assignments

No of Channels	Channel Assignment
3	1, 3, 6
4	1, 3, 6, 10
5	1, 3, 7, 10, 15
6	1, 4, 8, 13, 19, 21
7	1, 3, 11, 17, 22, 26, 29
8	1, 4, 9, 15, 19, 31, 38, 40
13	1, 3, 6, 10, 23, 31, 42, 54, 69, 87, 111, 121, 127

The spread of odd order intermod products is shown in Fig. 4.5 This assumes an allocated band BHz wide for the base transmitter and assumes that the actual assignments at any one site can be anywhere within this band. On this basis the bands shown are blighted by the intermod band and have a spread of nB Hz. For any blanket planning considerations therefore, the associated receiver band at the same site should be clear of the intermod spread up to the order which gives troublesome product levels. For 50W transmitters, measurements have shown that this should be at least the 11th order.

It is in principle possible to seek receiver assignments within the blighted bands on the assumption that the intermods will be discrete rather than continuous. This becomes difficult for moderate numbers of channels due to the number of products (see the next section), the chore of calculating and assessing them (even with processor help), and the impact of modulation. For these reasons the simplest approach is to cluster the actually assigned transmit frequencies as closely as possible to form a sub band B' Hz wide such that B' is smaller than B and the subsequent intermod spread is correspondingly reduced. Thus a localised assignment band B' is formed which can be located anywhere within B.

The arguments used in this section have assumed that the designer has only two bands of frequencies to contend with (one transmit, one receive) and that he has full control of the site. It is not uncommon, however, to find that there are other users, or site sharers, present. Just one other user transmitting outside the band B can cause the product distribution to be as shown in Fig.4.6. The extra bands which are blighted are shown shaded. Thus just one extra radiated signal can significantly exacerbate the frequency assignment problems where intermods have to be taken into account.

4.4.8 Number of In-Band Products

It is instructive to gain an idea of the rate at which the number of intermod products increase as extra numbers of channels are considered and/or higher order numbers. The in-band (i.e. odd order numbers with a difference of 1 in the number of positive to negative contributions) case is considered since it gives an idea of the difficulty of finding

TABLE 4.5 Number of in-band Intermodulation Products

Nonlinearity Number order channels n	3rd	5th	7th	9th
2	2	2	2	2
3	9	15	21	27
4	24	64	124	204
5	50	200	525	1095
6	90	510	1770	4626
7	147	1127		

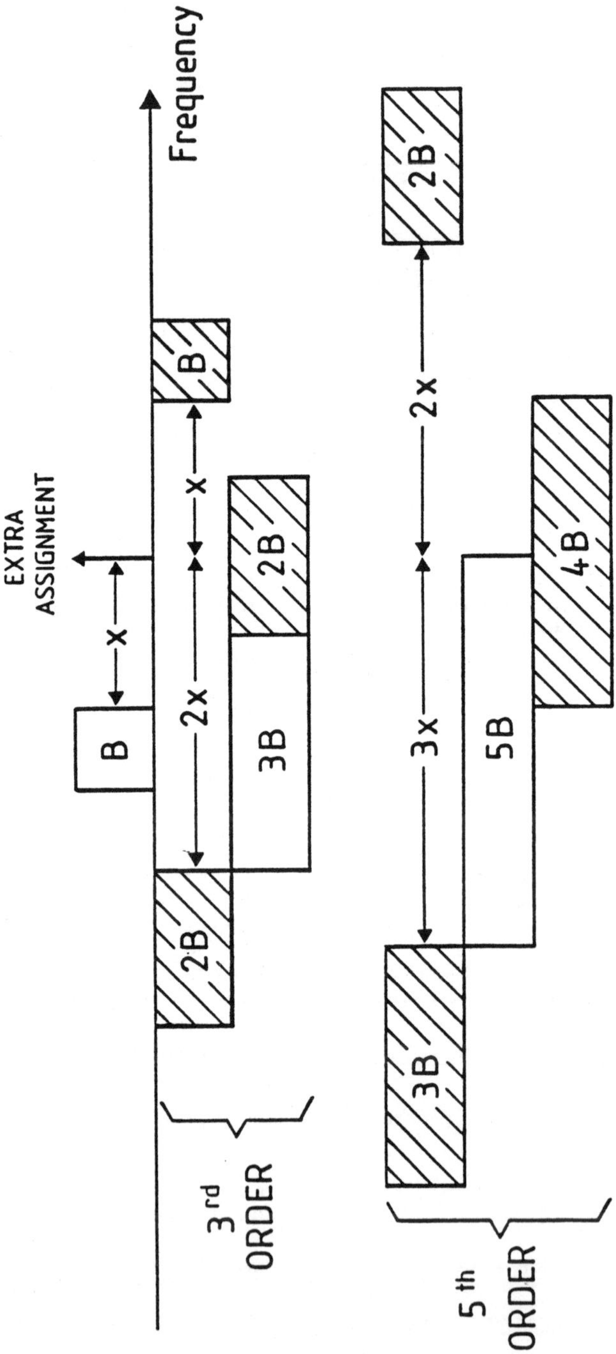
Frequency
EXTRA ASSIGNMENT
x
B
2x
x
B
2B
3B
2B
3rd ORDER
2x
2B
3x
5B
4B
3B
5th ORDER

Fig. 4.6

intermod free locations for the situation considered in the previous sections. These are shown in Table 4.5.

4.5 PREDICTING INTERMODULATION PRODUCT AMPLITUDES

So far the analysis has demonstrated how the location of intermodulation products in frequency may be determined, but the problem remains of predicting their magnitudes. Two aspects are immediately apparent:

1) Can an analytical model be found for the processes which take place at a real site which will be tractable mathematically whilst still retaining the essential features of the practical system?
2) Having found such a model, can its parameters be determined by experiments at "real" frequencies?

The need for a nonlinear relationship between an input function and an output function is immediately evident and again a choice exists between two possibilities, a) a nonlinear immittance in which the voltage and current in a single port network are nonlinearly related, or, b) a nonlinear two-port in which input and output voltages or currents are related by a nonlinear transfer function. The second alternative is more attractive as a basis for computation since the hypothetical two-port may be assumed to have linear input and output impedances with the intermodulation generating mechanism confined to a nonlinear transfer function. This simplifies the problem of specifying the various inputs to the network in terms of transmitter output powers, output impedances and mutual isolations. Formulation of the transfer function is clearly the next step.

4.5.1 General methods

4.5.1.1 Representation by Volterra or Wiener Functional Series

Returning to the polynomial transfer characteristic of equation (4.1), it is necessary to consider what form the coefficients A_1 - A_n must take to adequately represent the practical nonlinear interaction processes. If these coefficients are assumed to be real (i.e. non complex) and constant, then a number of restrictions are imposed, in particular AM to PM conversion processes are excluded and the process becomes entirely memoryless. However, a completely general representation of any nonlinear system may be obtained by replacing each term in the polynomial of (4.1) by a multi-dimensional convolution integral. The result, known as a Volterra series, takes account of the history of previous inputs to the general nonlinear system in addition to the instantaneous response and includes AM-PM conversion mechanisms characteristic of nonlinear reactive components in the system. The series now takes the form

$$y(t) = \int_{-\infty}^{\infty} h_1(\tau_1)x(t-\tau_1)d_1 + \int_{-\infty}^{\infty}\int_{-\infty}^{\infty} h_2(\tau_1,\tau_2)x(t-\tau_1)$$

$$x(t-\tau_2)d\tau_1 d\tau_2 + \int_{-\infty}^{\infty}\cdots\int_{-\infty}^{\infty} h_n(\tau_1,\tau_2\ldots\tau_n)$$

$$x(t-\tau_1)\ldots\ x(t-\tau_n)d\tau_1\ldots\ d\tau_n \qquad (4.4)$$

$h_n(\tau_1,\tau_2\ldots\tau_n)$ is referred to as the nth order Volterra kernel. Each term may be regarded as a subsystem in parallel with every other specified by an impulse response $h_n(t_1\ldots t_n)$. As in the linear case, integral transform procedures may be followed to simplify solution in real systems and a multi-dimensional transform pair may be defined as

$$H_n(j\omega_1,j\omega_2\ldots j\omega_n) = \int\ldots\int\ h_n(\tau_1,\tau_2\ldots\tau_n)\ x$$

$$\exp -j(\omega_1\tau_1+\omega_2\tau_2\ldots)\ d\tau_1 d\tau_2\ldots\ d\tau_n \qquad (4.5)$$

with the usual inverse (i.e. with $1/(2\pi)^2$ preceeding the RHS integral).

The function $H_n(j\omega_1,j\omega_2\ldots)$ is called the nth order kernel transform.

In treating a practical circuit by this technique, the circuit elements may be described by individual nonlinear current/voltage relationships, the kernels evaluated and the system response determined. Two problems are apparent, however:

1) evaluation of the kernels is clearly a complex operation and in consequence the technique is only really practicable provided the nonlinearities are sufficiently mild (or weak) for convergence of the series to be obtained in at most fifth-order terms,
2) if the system is regarded simply as a black-box with unknown internal configuration, identification of the Volterra kernels must rely on direct measurements at the network terminals.

In the case of linear systems 2) would be achieved by examining the impulse response of the network but in the generalised nonlinear case, this impulse response becomes multi-dimensional. From a purely mathematical standpoint, techniques may be defined for obtaining such responses (10) but translating these procedures into practical tests at realistic frequencies presents formidable problems which so far remain unsolved.

The extension to the Volterra approach due to Wiener(11) utilises a procedure by which a set of orthogonal functions are formed from the Volterra functionals under the constraint that orthogonality exists when the input is a white Gaussian time function. In general system theory terms, this technique has the advantages that it a) extends the range of nonlinearities over which converge nce can be obtained and b) offers the possibility that the orthogonal functionals may be determined by the system response to white noise rather than sets of time related impulses. Nevertheless, so far this procedure has only been applied in practice to physiological systems with very low frequency responses. Extension to VHF and UHF does not appear practicable at present.

4.5.1.2 The generalised polynomial representation

An alternative to the foregoing procedure might be to preserve the form of equation (4.1) but to generalise the coefficients $A_1, \ldots A_n$. To be completely general, the A_n must be capable of including phase information and the amplitude to phase transfer mechanism and this implies that the A_n must be complex quantities. Additionally, frequency dependent time delay is a property of a broad class of frequency dependent nonlinear systems and must be incorporated. Consequently, the process of computing the output spectrum obtained when the hypothetical network is stimulated by a multi-frequency input is by no means straightforward. Steer and Kahn(9) have demonstrated that in a system in which individual nonlinearities in components are known and expressable separately as power series in applied current or voltage, satisfactory computation schemes can be demonstrated so the first criterion referred to at the beginning of this section is met. However, if the circuit elements are not individually defined, the problem of identifying an equivalent model and obtaining the various parameters required from external measurement appears for the present at least to be impracticable unless simplifying assumptions can be made.

4.5.2 Approximate methods

The solution to these difficulties must lie in approximation and several techniques have been advocated.

4.5.2.1 Polynomial representation with real coefficients

This is easily the most readily accessible procedure in which the coefficients of equation (4.1) are assumed to be real and constant. Components in the output function are then obtained (12) as

$$y_N = e_N \sum_{i=1}^{M} |x_1|^{|k_i|} \sum_{L=o}^{\infty} \frac{A_{N+2L}(N+2L)!}{2^{N+2L-1}} \sum_{q_1+q_2 \cdots q_M=L} \prod_{p=1}^{M} \frac{|x_p|^{2q_p}}{q_p!(|K_p|+q_p)!} \quad (4.6)$$

for M input signals and y_N is the amplitude of a product at $(k_1\omega_1+k_2\omega_2 \ldots)$ with $N=|k_1|+|k_2|+ \ldots$ $e_N = 1$ for $N = 0$ and 2 for $N = 1, 2, \ldots$

The attractions of this method are immediately evident in that product magnitudes are determined from summations and product functions with integer increments rather than continuous integrals so that computation is rapid and the values of the coefficients A_n can be directly determined by measurements on the output spectrum resulting from a two-tone input, since the relationship between the A_n and the magnitudes of individual products can be explicitly defined (13,14). However, this approximation, although useful in many cases, is suspect when accuracy is of first importance. This is because in allowing the A_n to take real, frequency- and time-invariant values features of practical system characteristics are lost. As indicated earlier, the model is memoryless and AM-PM conversion processes cannot be taken into account. All other approximate techniques aim to circumvent these shortcomings by various strategies concentrating particularly on the AM-PM conversion issue which is very significant in the co-sited transmitter case because of the nonlinear reactive component of the output impedance which is characteristic of both tube and solid state transmitter output stages.

4.5.2.2 Amplitude-phase and quadrature models

The starting point of amplitude-phase models is the representation of the output by a function y(t) given by

$$y(t) = A[r(t)] \cos\{\omega_0 t+\chi(t) + \Phi[r(t)]\} \tag{4.7}$$

where r(t) is the envelope function and $\chi(t)$ the phase function associated with an input signal given by

$$x(t) = r(t) \cos [\omega_0 t + \chi(t)] \tag{4.8}$$

Advocates of this approach have proposed various functions for the A[r(t)] and Φ[r(t)] terms in equation (4.7) to give the most convenient and/or accurate representation of the particular nonlinear system under study (the TWT satellite amplifier has received much attention (15)).

The quadrature model of Fig. 4.7 is related to the amplitude phase models since if

$$P[r(t)] = A[r(t)] \cos \{\Phi[r(t)]\}$$

$$Q[r(t)] = A[r(t)] \sin \{\Phi[r(t)]\} \tag{4.9}$$

then in Fig.4.7

$$p(t) = P[r(t)] \cos [\omega_0 t + \chi(t)]$$

$$q(t) = - Q[r(t)] \sin [\omega_0 t + \chi(t)] \tag{4.10}$$

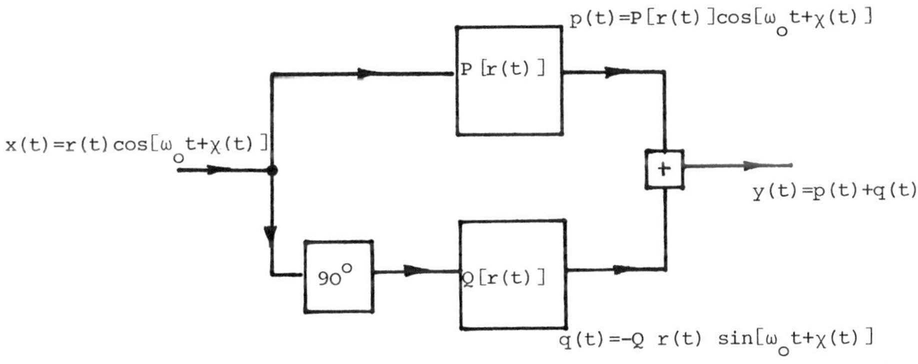

Fig. 4.7 Quadrature model after Saleh (ref. 15)

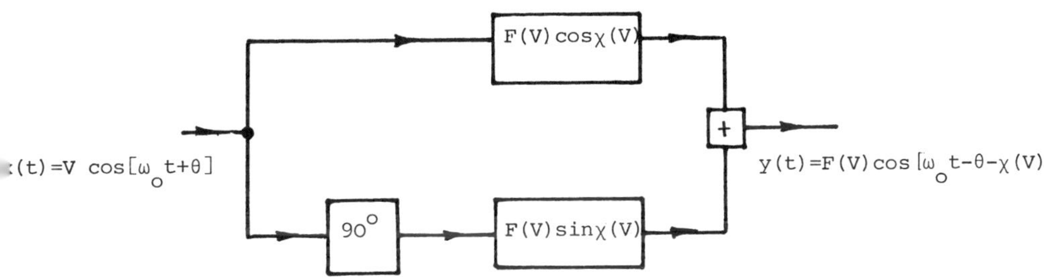

Fig. 4.8 Quadrature model after Abuelma'atti (ref. 17)

Again the quadrature model has attracted much attention and several approaches have been used to obtain predictions of intermodulation spectra under different conditions of input signal. A general result for the output function may be obtained for a multi-component input by setting

$$x(t) = \sum_{i=1}^{n} V_i \cos[\omega_i t + \chi_i(t)] \tag{4.11}$$

$$y(t) = \sum_{\substack{k_1,k_2\ldots k_n = -\infty \\ k_1+k_2\ldots k_n = 1}}^{\infty} M(k_1,k_2\ldots k_n) \times \exp\left\{\sum_{i=1}^{n} i[k_i\omega_i t + k_i\chi_i(t)]\right\} \tag{4.12}$$

Where $M(k_1,k_2\ldots)$ is the complex amplitude of a product of frequency $k_1\omega_1+k_2\omega_2+\ldots$ given by

$$M(k_1,k_2\ldots k_n) = \int_0^{\infty} s\left[\prod_{i=1}^{n} K_{j_i}(V_i s)\right]ds \times \int_0^{\infty} r[P(r) + j\,Q(r)]\,J_1(rs)\,dr \tag{4.13}$$

The transmitter modelling problem now reduces to one of finding appropriate functions for P(r) and Q(r) such that the parameters of the function are determinable from measurements. Saleh (15) recommends particularly convenient two-parameter functions for modelling TWT's, i.e.

$$P(r) = \alpha_p r/(1+\beta_p r^2)$$

$$Q(r) = \alpha_q r^3/(1+\beta_q r^2)^2 \tag{4.14}$$

where the four parameters $\alpha_p, \beta_p, \alpha_q$ and β_q are sufficient to define the device under study. However, these functions are not generally applicable to the transmitter modelling problem where amplitudes and phase functions may take arbitrary form.

An alternative has been used with success by Abuelma'atti and Gardiner (16) which relies on sine function approximations to the quadrature model path characteristics. This involves representation of the instantaneous transfer function of the nonlinear network by a Fourier series in instantaneous input voltage with periodicity chosen such that the working range of the practical peak input voltage is an appropriate segment around zero. If the Fourier series comprises only sine functions, then this segment clearly passes through the origin, i.e.

$$V_o(t) = \sum_{m=1}^{M} \gamma_m \sin \frac{m\pi}{B} V_i(t) \qquad -B < V_i(t) < B \tag{4.15}$$

where

$$\gamma_m = \frac{1}{B} \int_{-B}^{B} V_o(t) \sin \frac{m\pi}{B} V_i(t) \; d\, V_i(t) \tag{4.16}$$

with M being the number of sine functions required to achieve adequate representational accuracy, 2B is the repetition period with $V_i(t)$ and $V_o(t)$ input and output instantaneous voltages respectively. The quadrature model then takes the form of Fig. 4.8.

If an input signal comprising several components is applied, i.e.

$$V_i(t) = \sum_{k=1}^{K} V_k \cos \omega_k t \tag{4.17}$$

Then intermodulation product amplitudes are derived as

$$V_{\alpha_1,\alpha_2 \cdots \cdot \alpha_k} = 2 \sum_{m=1}^{M} \gamma_m \prod_{k=1}^{K} J_{|\alpha_k|}\left(\frac{m\pi}{B} V_k\right) \tag{4.18}$$

where $V_{\alpha_1,\alpha_2 \cdots \alpha_k}$ refers to the product at frequency $\alpha_1 f_1 + \alpha_2 f_2 \cdots \cdot \alpha_k f_k$ and the α_k are positive or negative integers of zero and

$$\sum_{k=1}^{K} |\alpha_k| \quad \text{is an odd integer.}$$

For a device exhibiting AM/PM conversion the γ_m become complex but by representing the nonlinearity by two parallel paths in quadrature, as shown in Fig. 4.8, the γ_m may be replaced by two real quantities, δ_m and β_m, for in-phase and quadrature characteristics respectively and intermodulation product amplitudes computed for each path separately. This gives

$$V_{(\alpha_1,\alpha_2 \cdots \cdot \alpha_k)p} = 2 \sum_{m=1}^{M} \delta_m \phi(|\alpha_k|, V_k) \tag{4.19}$$

$$V_{(\alpha_1,\alpha_2 \cdots \alpha_k)q} = 2 \sum_{m=1}^{M} \beta_m \phi(|\alpha_k|, V_k)$$

where suffices p and q refer to in-phase and quadrature paths respectively and

$$\phi(|\alpha_k|, V_k) = \prod_{k=1}^{K} J_{|\alpha_k|}\left(\frac{m\pi}{B} . V_k\right) \tag{4.20}$$

This gives finally

$$V_{\alpha_1,\alpha_2\ldots\alpha_k} = \sqrt{V_{(\ldots\alpha_k\ldots)}{}^2_p + V_{(\ldots\alpha_k\ldots)}{}^2_q} \qquad (4.21)$$

4.5.2.3 Incorporating frequency dependence and obtaining the model parameters

The foregoing models remain memoryless in that there is no frequency dependence built into the functions which define the nonlinear characteristics. To represent the practical system therefore passive linear frequency dependent networks must be added. It is usually considered adequate to embed the nonlinear device between filter networks at input and output and indeed it may be shown that in the case of the quadrature model with nonlinear functions represented by polynomials a Volterra series representing the complete system may be derived which then has memory properties. Saleh (15) on the other hand, recommends incorporating frequency dependence in each path of his quadrature model by making the α and β parameters frequency dependent. Some doubt remains regarding the relative merits of external or built-in frequency dependence which is the subject of current theoretical study. However, recent experiments (17) have indicated that the system model of a communal site described in ref. 13 with the quadrature nonlinear representation of ref. 16 gives an optimum compromise between accuracy and ease of measurement.

Intermodulation product amplitudes and phase information are therefore readily obtainable and the remaining problem is the determination of the δ_m and β_m from measurements. Ideally, if product amplitudes and phases could be directly measured, then the model parameters are immediately available by matrix inversion but the measurement of product phase presents almost insurmountable difficulties. Abuelma'atti (17) has advocated a surface fitting procedure based on measurements of 3rd order product magnitude alone, but whilst this appears to have given satisfactory results in some cases its general application is suspect on grounds of accuracy when very wide dynamic ranges are involved.

Approaching the problem from a different standpoint, it is seen that if enough product amplitude information is available then selection of appropriate δ_m and β_m becomes a quadratic optimization process because the relative magnitudes of $V_{(\alpha_1,\ldots\alpha_k\ldots\alpha_K)_p}$ and $V_{(\alpha_1\ldots\alpha_k\ldots\alpha_K)_q}$ are unknown but the $V_{(\alpha_1\ldots)_p}^2 + V_{(\alpha_1\ldots)_q}^2$ are measurable.

Currently the difficulty is overcome by making the further simplifying assumption that while fundamental signal phases are amplitude dependent, intermodulation product phases are not. This simplification is significant because

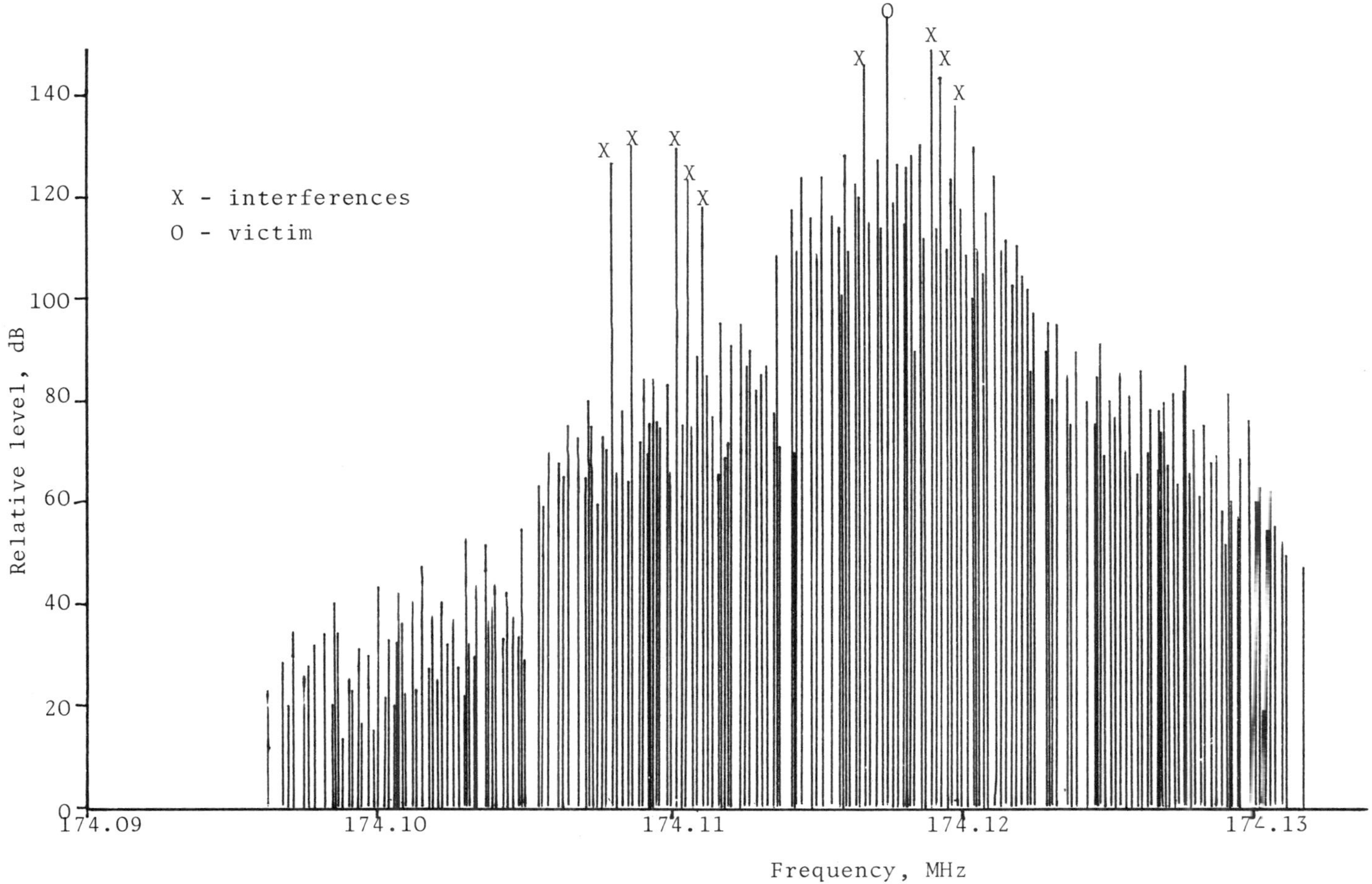

Fig. 4.9 Computed spectrum of third order intermodulation generated by 10 interacting transmitters

it permits the δ_m and β_m to be determined by matrix inversion using measurements of fundamental amplitude and phase and product amplitudes alone. By way of conclusion, Fig. 4.9 shows a typical computed result for the spectrum generated by a number of mutually interacting transmitters.

REFERENCES

1 ITT Reference Data for Radio Engineers, Harold W Sams & Co Inc, 1975

2 Connor, F R: 'Antennas', Edward Arnold, 1974

3 Balanis, C A: 'Antenna Theory, Analysis and Design', Harper & Row, New York, 1982

4 Bostick, G: 'A set of practical design data for quarter-wave coaxial (TEM) cavities', Microwave Journal, p.78, April 1960

5 Gardiner, J G & Mgombelo, H R: 'Hybrid dividers for VHF and UHF Systems', IEE Conf Radio Receivers & Associated Systems', Leeds 1981

6 Howson, D P: 'The basic principles of VHF & UHF filter duplexers', Postgraduate School of Electrical & Electronic Engineering, University of Bradford, Research Report No 246

7 Davidson, Allen L: 'Omnidirectional transmitter combining antenna' IEEE Trans on Vehicular Technology, Vol VT 32, No 2, May 1983

8 Edwards, R, Durkin, J & Green, D H: 'Selection of interference-free frequencies for multiple channel mobile radio systems', Proc. IEE, Vol 116, No 8, August 1969

9 Steer, M B and Khan P J: 'An algebraic formula for the output of a system with large-signal, multifrequency excitation, IEEE Proc. Vol 71, No 1, 1983

10 Schetzen, M: 'The Volterra and Wiener theories of non-linear systems', John Wiley & Sons, 1980

11 Schetzen, M: 'Nonlinear system modelling based on the Wiener theory', Proc IEEE, Vol 69, No 12, December 1981

12 Sea, R G: 'An algebraic formula for amplitudes of intermodulation products involving an arbitrary number of frequencies', Proc IEEE, August 1968, pp 1388-1389

13 Al Hafid, H T, Atek, K, Abbas, M A H & Gardiner, J G: 'Computer modelling of nonlinear interactions between co-sited transmitters', IERE Conf on Civil Land Mobile Radio, NPL, November 1975

14 Al Hafid, H T & Gardiner, J G: 'The characterisation of VHF radio transmitter output nonlinearity for communal site spectrum prediction', Univ. of Bradford, Research Report No 204, Dec. 1975

15 Saleh, A A: 'Frequency-independent and frequency-dependent nonlinear models of TWT amplifiers', IEEE Trans on Communications, Vol COM-29, No 11, Nov 1981

16 Abuelma'atti, M T & Gardiner, J G: 'Input power assignment of equal-amplitude multicarrier systems for given output carrier power', Int. J. Electronics, Vol 50, No 1, 1981, pp 55-60

17 M T Abuelma'atti: "On the characterization of memoryless bandpass non-linear systems from intermodulation measurements", Int.J.Electronics, Vol 54, No 5, 683-687 1983

Chapter 5

Modulation techniques

S.W. Watkinson

5.1 INTRODUCTION

This chapter describes the methods of modulation used in mobile radio. It covers the theory and some of the practical implications of that theory.

The intention is not only to derive equations that give correct answers, but also to describe the meaning of these theoretical explanations.

5.2 MOBILE RADIO FREQUENCIES

In land mobile communications the transfer of information is by modulating, transmitting and demodulating radio waves. The frequencies used are in the range 2MHz to approximately 1GHz, which corresponds to wavelengths of 150 metres to 30 centimetres. Generally, the lower frequencies give greater communication ranges than the higher frequencies. With the higher frequencies a line of sight path must normally exist between the transmitting and receiving aerials, or at least a path via reflective objects.

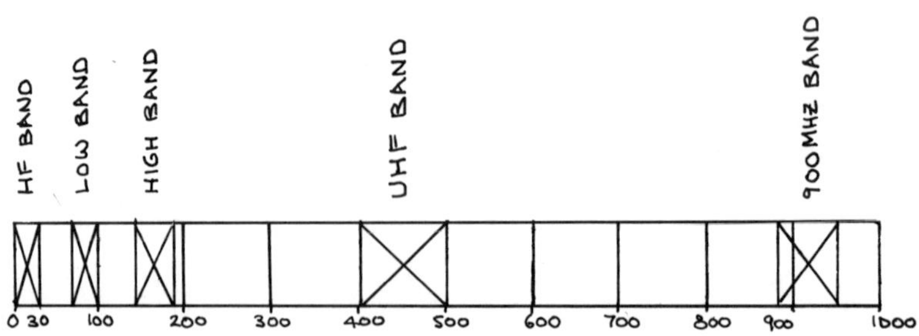

Fig. 2.1 Division of frequency spectrum for mobile radio.

Fig. 2.1 shows the approximate division of the frequency bands used in land mobile communications. The frequencies shown are approximate because individual countries have apportioned the frequencies used for land mobile and public broadcasting differently. The detail division in the 2MHz to 30MHz band is not shown.

The frequency spectrum is used by dividing it into channels, each channel being allocated to a particular user. The channels currently in use are between 30kHz and 12.5kHz wide. The bandwidth of the modulating signal sets a lower limit on the channel width. A speech signal is normally considered as having components between 300Hz and 3kHz, but a channel width of only 2.7kHz would be impractical. This is because the R.F. channel signal may drift slightly with temperature changes and distortion products will increase the signal bandwidth.

In addition the narrowband crystal filter defining the receiver channel width will have finite attenuation slopes. This leads to an apparent widening of the channel as the level of an adjacent interfering signal increases. When these factors are considered, the narrowest channel width is 5kHz if mutual interference between adjacent channels is to be kept at acceptable levels.

As there is considerable redundancy in a speech signal, it is possible to use smaller channel widths by employing complex circuits to analyse and synthesise speech signals. These techniques are beyond the scope of this chapter and are not yet in general use.

5.3 MODULATION

The information to be transmitted is contained in the base band signal, however it is not feasible to transmit it in this form. It is possible to produce electro magnetic radiation at base band frequencies, but an effective aerial would have dimensions of the order of miles, which is not practical for mobile radio.

A further problem would arise if all signals were transmitted at base band frequency, there would be no way of sorting out which signal belonged to a particular user.

To overcome these difficulties the process of modulation is used to transfer the base band signal to a particular allocated channel in the R.F. spectrum.

In very general terms, modulation is the process of changing something in a controlled manner. In radio communications it implies the variation of an R.F. carrier signal by the base band signal. Either the amplitude or frequency of the carrier can be varied, or a combination of the two.

By this means the information contained in a base band signal is transferred to a particular narrow slot in the R.F. spectrum. If a receiver is tuned to that same narrow band of R.F. frequencies, it will pick out the correct required signal. Much of the cost and complexity in a receiver is devoted to defining this narrow band of frequencies and constitutes the receiver selectivity.

5.4 AMPLITUDE MODULATION

Before radio was invented, a very basic form of modulation was being used in telephones. A carbon microphone was used to vary the d.c. current flowing through the circuit. When the first radio transmitters were developed, they produced radio waves covering a wide range of frequencies using a high voltage spark gap. These waves conveyed information by being turned on and off; an extreme form of amplitude modulation. When it became possible to produce radio signals on one frequency, this was again switched on and off, according to the Morse Code, to convey a message.

It is very likely that radio operators listening to morse noticed the whine of the electric generator and realised that the amplitude of the transmitted waves could also be varied by a speech signal from a microphone. In this way Amplitude Modulation (AM) was born. A later section on practical modulator techniques will show that AM can be generated, as would be expected, by varying the amplitude of a carrier. However an alternative method of generating AM is shown in Fig. 4.1 which leads to a better understanding of the concept of side bands. It will be shown that the amplitude of the output signal is directly proportional to the modulating base band waveform. This approach also assists the understanding of Single Side Band (SSB) and Double Side Band Suppressed Carrier (DSBSC). A carrier is used to generate these signals, but is not present in the output of the transmitter.

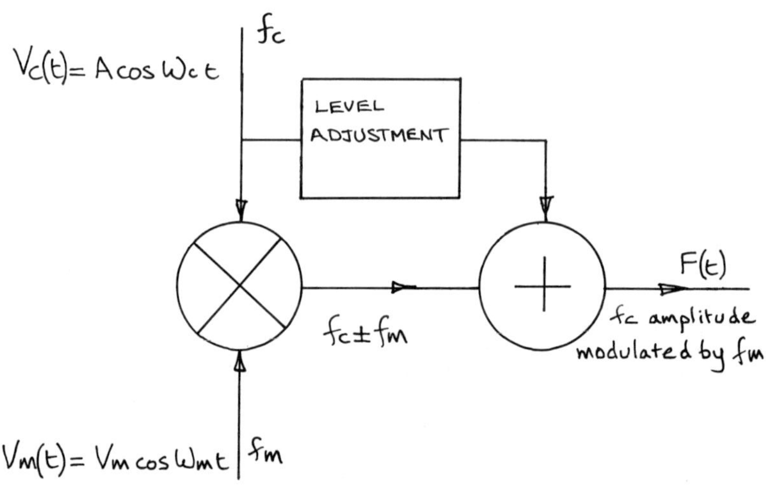

Fig. 4.1 Simple amplitude modulator

Amplitude modulating a carrier with an audio waveform is described in mathematical terms as multiplying them together and adding the unmodulated carrier as shown in Fig. 4.1.

Multiplying is generally called mixing and it is accepted that the output of a mixer consists of the sum and difference frequencies of the inputs. There are other lower level outputs, but these two are the main ones.

If the carrier frequency is f_c

and the modulating frequency f_m

then the output of the mixer is $f_c \pm f_m$

If unmodulated carrier is added to the mixer output the result will be as in Fig. 4.2.

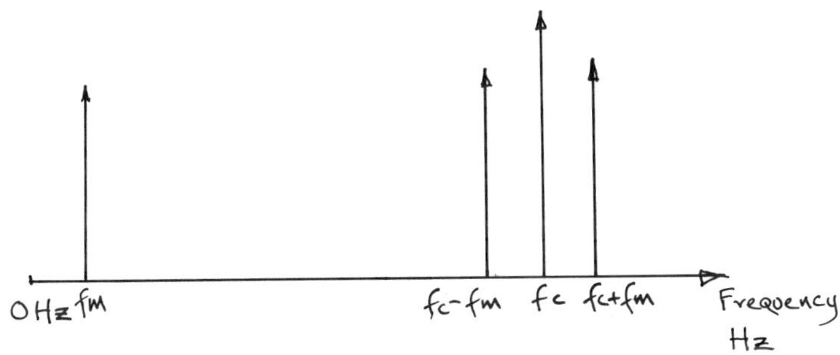

Fig. 4.2 Spectrum of amplitude modulation.

If the modulating signal consists of a band of frequencies, for instance an audio waveform from a microphone, then as in Fig. 4.3 the amplitude modulator carrier will have bands of frequencies above and below f_c. These are called the upper and lower sidebands.

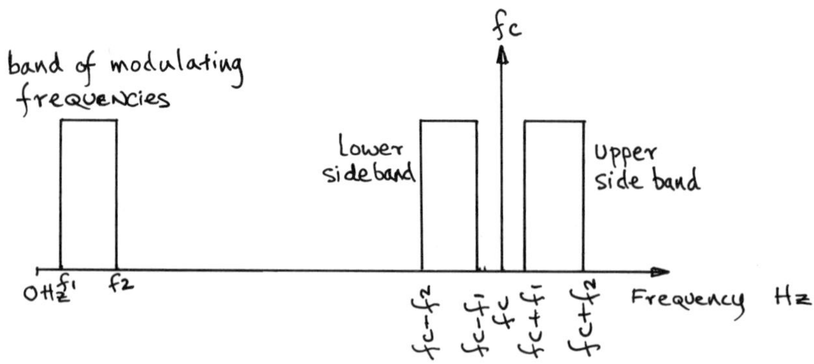

Fig. 4.3 Illustration of side bands.

Some readers may consider that this approach to describing AM is rather complicated; that is when compared with the usual concept of varying the amplitude of a carrier in sympathy with the modulating signal. However such an approach does not make the existence of side bands very obvious.

In the early days of radio when the theory was less well understood, there was much argument about the reality of side bands. Particularly so when it was first realised that the total bandwidth of an AM signal was twice that of the modulating signal. These doubts have long since gone and a spectrum analyser will demonstrate that the theory is correct and the side bands are really there.

5.5 A CLOSER LOOK AT THE THEORY OF AM DSBSC AND SSB

5.5.1 The theory of Amplitude Modulation (AM)

Consider again the arrangement shown in Fig. 4.1 where

$$A\cos\omega_c t \quad \& \quad B\cos\omega_c t$$

are sinusoids at the RF channel frequency and

$$V_m \cos\omega_m t$$

is the modulating signal.

The output of the multiplier = $A \cos W_c t . \; V_m \cos W_m t$ ------ 5.1

Output of the adder = $F_{(t)}$

$$F(t) = A \cos W_c t . V_m \cos W_m t + B \cos W_c t \quad \text{-------- 5.2}$$

$$= (A V_m \cos W_m t + B) \cos W_c t \quad \text{------- 5.3}$$

$$= B\left(\frac{A V_m}{B} \cos W_m t + 1\right) \cos W_c t \quad \text{------ 5.4}$$

$$F t = B(m_a \cos W_m t + 1) \cos W_c t \quad \text{----- 5.5}$$

Where $m_a = \dfrac{A V_m}{B}$ ------ 5.6

is called the modulation index or modulation factor. B is the carrier voltage amplitude without modulation and A is the input carrier level to the multiplier.

The envelope of the modulated wave = $B(m_a \cos W_m t + 1)$ --- 5.7

Note that the deviation of the envelope from its unmodulated value is proportional at every instant to the value of the modulating signal. When expressed as a percentage, Ma is said to be the Percentage Modulation. The value of Ma should not exceed unity, or 100 percent, otherwise the negative envelope peaks will invert.

Fig. 5.1 shows the way the envelope of F(t) changes for different values of modulation factor Ma. For $M < 1$, the modulated carrier amplitude varies in proportion to (Ma Cos Wmt + 1). It is also worth noting that the modulated carrier frequency Wct is not changed by the modulation process.

It is particularly convenient that the envelope shape of the modulated wave contains the modulating information. This means that a simple diode rectifier can be used to recover the modulation by charging a capacitor to follow the envelope shape.

The spectrum of F(t) can be seen more readily by using the trig. identity.

$$2 \cos A \cos B = \cos(A + B) + \cos(A - B)$$

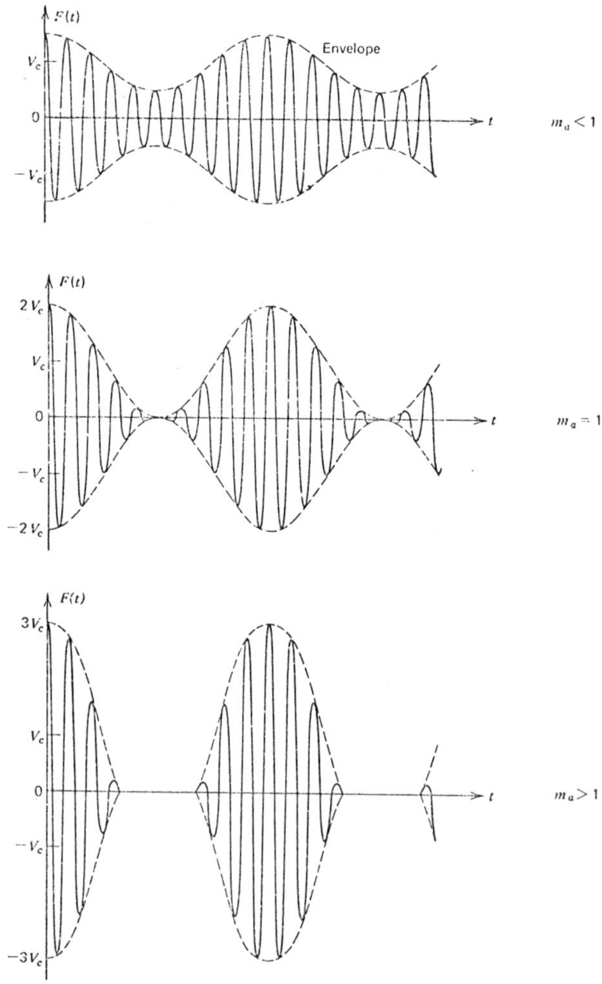

Fig. 5.1. Envelope shape of amplitude modulated wave.

to change the form of the expression:-

$$F(t) = B(ma \cos W_m t + 1) \cos W_c t \quad ----- 5.5$$

when rearranged this gives:-

$$F(t) = B \cos W_c t + \frac{Bma}{2} \cos(W_c + W_m)t + \frac{Bma}{2} \cos(W_c - W_m)t \quad --- ----- 5.8$$

the first term B Cos Wct is called the 'carrier' and is the component added by the summing network shown in Fig. 4.1. As would be expected, this term is not affected by the modulation process. The other two terms are the upper and lower side bands as described earlier.

The amplitude of these side band terms is proportional to the modulation index Ma, and their frequencies are:-

$$(W_c + W_m) \quad ----- 5.9 \text{ for the upper side band}$$

$$(W_c - W_m) \quad ----- 5.10 \text{ for the lower side band}$$

These are called the sum and difference frequencies of the carrier and modulation.

One set of side band frequencies will be a mirror image of the other, hence all the modulation information is contained in one side band. In addition the carrier does not contain any information and can in principle be filtered out prior to transmission, or simply not added.

The two sets of side bands without the carrier constitute a Double Side Band Suppressed Carrier (DSBSC) signal. Such a modulation system has been proposed for mobile radio use, however both the transmitter and receiver become more complex without any clear advantage.

If one side band and the carrier are omitted, the remaining side band can be transmitted as a Single Side Band (SSB). This arrangement has the advantage of using only half the RF bandwidth of an AM signal and may be used for mobile radio when all the available spectrum is filled with channels. More will be said later about the practical aspects of these different forms of modulation derived from AM.

5.5.2 Phasor representation of AM signals

Some of the properties of modulation systems can be illustrated by phasor diagrams as shown in Fig. 5.2. The carrier is shown by phasor Ac which rotates at the uniform velocity Wc. It is a convention that the observer rotates with Wc, it is therefore shown stationary.

To that observer the phasors that represent the side bands will rotate at the difference angular velocities. The phasors representing A.M. rotate in opposite directions but at equal angular velocities, hence the resultant of the carrier plus side bands remains in phase with the carrier. This confirms the previous observation that the frequency of an A.M. signal is unaffected by modulation.

It will be apparent that the resultant of DSBSC also has a constant angular velocity Wc. However this is not the case with SSB, in that case the angular velocity of the resultant depends upon both the amplitude and frequency of the modulating signal. These results agree with equation 5.8 derived earlier for the A.M. signal showing the carrier and two side bands.

5.5.3 Power in an A.M. signal

Referring again to equation 5.8 and assuming that the voltage is applied to a 1 ohm resistor, the average power of the carrier for sinusoidal modulation is:-

$$\frac{V_c^2}{2} \qquad \text{------} \; 5.11$$

Similarly the power in each side band is:-

$$\frac{V_c^2 m_a^2}{2} \qquad \text{------} \; 5.12$$

Hence as the percentage of modulation increases from 0 to 100% the total power in the carrier plus side bands increases from:-

$$\frac{V_c^2}{2} \quad \text{to} \quad \frac{1 \cdot 5\, V_c^2}{2}$$

This extra power has to be provided by the multiplier in the arrangement in Fig. 4.1.

5.6 SOME PRACTICAL ASPECTS OF A.M. TRANSMITTERS

5.6.1 Amplitude Modulation

This is relatively easy to produce; this is mainly because, as the theory shows, there is no information contained in the phase of the signal. Some of the popular amplitude modulation methods in use will now be described.

5.6.2 High level modulation of a Class C amplifier

A typical arrangement is shown in Fig. 6.1. This is the most efficient way to generate an A.M. signal and is the most widely used method in mobile and fixed transmitters. The modulation signal is used to vary the supply voltage to the R.F. amplifier between zero and twice the nominal supply voltage. Such a variation covers the range of 0 to 100% modulation. It is worth noting that at 100% modulation the peak envelope power is four times the unmodulated power. This means that the transmitter output stage must be rated at four times the carrier power rating.

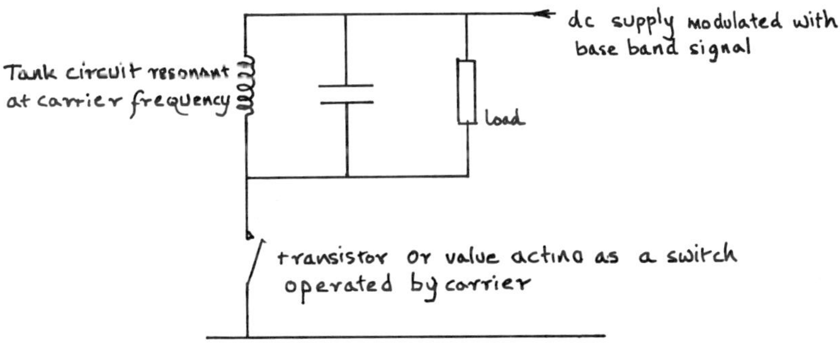

Fig. 6.1. Class C amplifier A.M. modulator

The efficiency of this method can be as high as 80%, this is because a Class C amplifier is very efficient at converting power supply energy into R.F. energy. In such an arrangement the modulator is acting as a multiplier; the modulating signal is added to a D.C. component, this adds in a carrier and so conforms with the theoretical approach to A.M.

5.6.3 Low level modulation of a Class B or C amplifier

In this method the modulating signal is used to vary the bias on the grid of a valve, the base of a bipolar transistor or the gate of a field effect transistor. The changing bias alters the R.F. conduction angle of the stage to vary the R.F. output level. Changing the conduction angle also varies the efficiency of the stage and it follows that such an arrangement will have a lower overall efficiency. If a large modulation depth is required, some form of linearisation must be applied to the modulation control point.

This form of low level amplitude modulator is useful when the modulating signal has a wide bandwidth, as it is not easy to design high power wide band modulators.

5.6.4 Low level modulation followed by linear power amplification

This technique uses a balanced mixer followed by addition of the carrier as described earlier. This method was originally developed for SSB transmitters, the unwanted side band being removed by filtering. It has since become popular for A.M. with the widespread use of semi-conductor devices in R.F. power amplifiers. Such devices are difficult to modulate at high power levels due to the large change in junction capacitance as the voltage across the junction is varied. This leads to detuning and spurious phase modulation.

5.7 SOME PRACTICAL ASPECTS OF SINGLE SIDEBAND

5.7.1 SSB System considerations

It was shown earlier that the bandwidth of an SSB signal is the same as that of the modulating signal. Hence this modulation technique uses the narrowest possible bandwidth. It is used for almost all communication channels in the 2 - 30MHz band where its use is well proven. The equation that was derived for SSB indicates that the modulating information is carried in both the phase and amplitude of the signal. This means that the whole of the signal path must have a linear phase and amplitude response. This includes the transmitter, the propagation medium and the receiver. Under some circumstances the wave path may vary with very small changes in frequency, so that modulation sidebands arrive at the receiver out of phase, leading to distortion. Because SSB uses less bandwidth than A.M. it is less affected by this kind of selective fading.

5.7.2 Frequency Stability requirements for SSB

To generate or demodulate an SSB signal, a stable reference frequency is required. This reference is often called the carrier, however as it is not normally radiated at a high level, it is a misleading name. If the reference signal is totally suppressed at the transmitter, the SSB receiver must generate the correct reference frequency by estimation. Any error will result in a shift in frequency of the demodulated signal. For speech communication the maximum acceptable error is approximately 20Hz. This means that the total frequency drift of all the oscillators affecting the real or notional carriers in the transmitter and receiver must not exceed 20Hz for all environmental conditions. This is commercially viable at transmission frequencies up to 30MHz, but is not viable in the VHF and UHF bands at 68MHz and above.

5.7.2.1 Single Side Band with Pilot Tone

To combat this and other problems, recent research has proposed the use of a transmitted pilot signal. This pilot could either be the 'carrier', a tone or an other continuous signal in the audio passband transmitted at approximately one tenth the peak envelope power of the components due to the modulating signal. This means that the pilot power is not a significant proportion of the rated transmitter output. Because of this lower peak envelope power there is less potential for interferring with other users of the spectrum.

A receiver for SSB with a pilot must be able to separate the speech side frequencies from the pilot. The pilot can then be used to control the frequency of the notional demodulating carrier wave as seen by the receiver. In this way demodulated audio frequency shifts can be eliminated, or at least reduced to an acceptable level. In addition the pilot can be used to determine an amplitude reference in the receiver to maintain a near constant output.

With a pilot tone system, the frequency drift that can be tolerated depends on the radio channel spacing and the acceptable adjacent channel interference limits. Oscillators that exhibit a 'part per million' frequency stability are currently available at a commercially acceptable price. Using such oscillators in systems operating up to 200MHz gives a minimum channel spacing of approximately 5kHz. This is less than half the narrowest practical channel spacing for A.M., F.M. or P.M. which is 12.5kHz. Such SSB systems will become attractive as the pressure for more land mobile radio channels increases.

5.7.3 Low distortion amplifier requirements for SSB

Inadequate frequency stability has not been the only problem to delay the introduction of SSB into the VHF part of the spectrum. Distortion in the power amplifier of an SSB transmitter will generate side frequencies that lie outside the channel bandwidth. These intermodulation products must be at a level well below those attainable with conventional linear power amplifiers. Recent work has yielded an ingenious solution to this problem in the form of the Polar Loop Transmitter. This system uses a combination of amplitude, frequency, and phase feedback techniques to achieve a low distortion high efficiency R.F. amplifier.

5.7.4 GENERATION OF THE SSB SIGNAL

There are three main methods of generating a low level SSB signal prior to amplification, these are described in the following sections.

5.7.4.1 The filter method

The theory of this method has already been discussed in Section 5.1. Double sideband suppressed carrier is first produced using a balanced mixer as a multiplier. The required sideband is filtered out and amplified in a linear amplifier to the required power level. It is usual to perform this mixing and filtering process at a fixed frequency below about 10MHz. This signal is then further mixed up to the required channel frequency. An advantage of the Polar Loop system is that this frequency translation is performed without generating further spurious products.

5.7.4.2 The Phasing Method

In this technique the 'carrier' and modulating signals are both applied to the two inputs of two balanced modulators. Each signal is applied to the two modulators via phase splitting circuits. These circuits maintain their two outputs at 90° apart over their operational frequency ranges, that is the 'carrier' range and the audio bandwidth. This has the effect of inverting the phase of the upper and lower sidebands from one of the mixers relative to the other. The outputs of the two modulators are added together, leading to the cancellation of one of the side bands and the re-inforcement of the other. This method entails maintaining a 90° phase shift between 300Hz and 3kHz to better than 0.5°, a difficult task. The technique works but is not very practical in a commercial equipment.

5.7.4.3 The Third Method

This is a further development of the phasing method requiring two more balanced modulators to overcome the need for wide band audio frequency phase shifters. It is rarely used commercially and because of its ingenious nature it is well described in the literature.

5.7.5 Single Side Band Receivers

The three methods of generating SSB can be used in reverse to demodulate the signal. The filter method is normally used because it is best suited to standard superhet radio receiver architecture.

5.8 ANGLE MODULATION

5.8.1 Basic concept of Angle Modulation

This form of modulation is used extensively and includes both frequency and phase modulation. The main distinguishing feature is the constant envelope amplitude of the radiated signal. It is the frequency that is varied by the modulating signal. Like AM and SSB this is a simple concept, however the theoretical analysis of the side bands generated is complicated. This is particularly so where the angular deviation due to modulation is high. In the system used for mobile radio the deviation is low and consequently the side band pattern is simplified.

5.8.2 Phasor representation of Angle Modulation

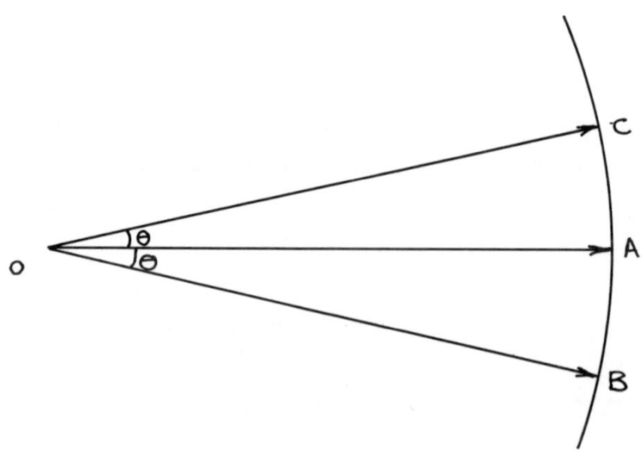

Fig. 8.1 Phasor diagram of Angle Modulated Signal

Fig. 8.1 is a phasor representation of an angle modulated signal. OA is the unmodulated carrier which rotates around O at a rate equal to its frequency. For convenience it is shown stationary, as if frozen by a stroboscope.

When the carrier is angle modulated it swings over a range of $\pm\ \theta$ between OA and OC.

The carrier could be caused to deviate by modulating the source oscillator, or using a separate phase modulator to introduce a variable phase shift. Such arrangements are used, but an alternative method enables the side band structure of an angle modulated signal to be better understood.

5.8.3 The Armstrong method of Angle Modulation

A very similar arrangement to that discussed in the AM section is shown in Fig. 8.2. The difference being the introduction of a 90° phase shift in the added carrier path. This phase shift has the effect of rotating the carrier relative to the counter rotating modulation vectors. It is a system devised by Armstrong in the early days of FM broadcasting.

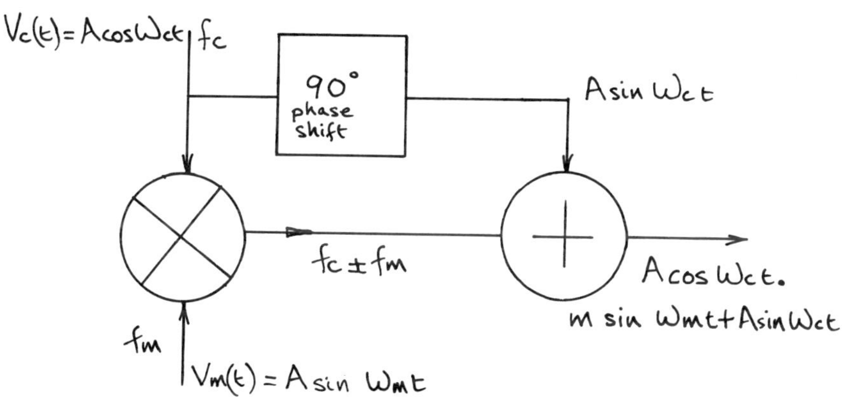

Fig. 8.2 The Armstrong method of Angle Modulation

The phasor diagram for this system is shown in Fig. 8.3.

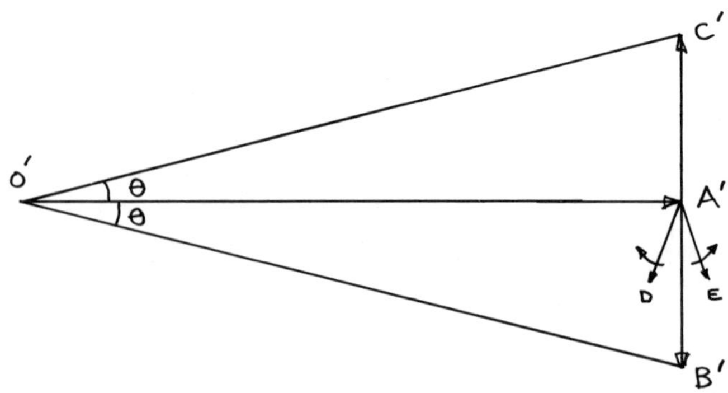

Fig. 8.3 Phasor diagram for the Armstrong method.

In this diagram AD and AE are the counter rotating modulation vectors, the resultant of these vectors always lies on the line BC. Notice that this resultant due to modulation is now in quadrature to the carrier O'A' and not in phase as it was in the AM case.

5.8.4 Shortcomings of the Armstrong method

This new phasor diagram is similar to that shown for true angle modulation in Fig. 8.1; the main difference is that the resultant of the modulation sideband B'C' no longer lies on the circumference of a circle. Because of this difference the amplitude of the resultant signal that swings between O'B' and O'C' is not of constant amplitude.

What further sidebands would be needed to maintain the signal amplitude constant? The answer is complicated in detail, but simple in concept. As the phasor O'A' swings between B' and C' it must be lengthened in the centre position and shortened at the limits of its swing.

5.8.5 How the Armstrong method can be improved

The shortcomings can be partially overcome by adding a further pair of rotating phasors such that their resultant is in phase with the carrier as shown below. These phasors must rotate at twice the modulation frequency, that is the second harmonic.

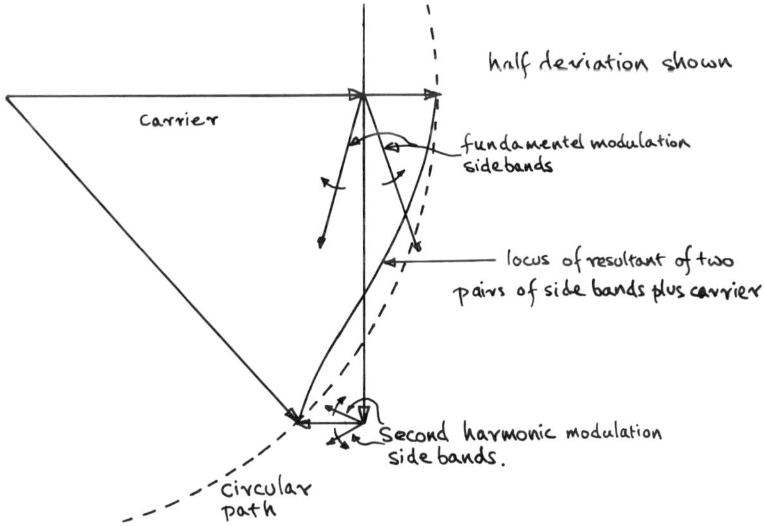

Fig. 8.4 Results of adding a second pair of sidebands

The diagram in Fig. 8.4 shows that the resultant phasor is the correct length at the extremes and centre of the angular deviation, but deviates from the circular path at all other points.

The complete answer is to add further pairs of counter rotating phasors at all the harmonics of the modulating frequency, the phasors of each subsequent pair being rotated by 90°. The relative magnitude of these numerous pairs of phasors will change according to the peak angular deviation required. It is possible, but complicated to draw a number of these additional vectors and show that they add and move inwards towards the correct point in a square sided spiral. This ensures that the resultant of the vector as it swings through the angular deviation always lies on a circle.

Fortunately mathematics comes to the rescue with a series which describes the relative amplitude of these phasors or sidebands. This series is called the Bessel Function and the diagram in Fig. 8.5 shows these terms plotted on a graph. It is of interest to note that the sidebands or the carrier have zero amplitude for certain values of angular deviation.

With the aid of a spectrum analyser, this disappearing carrier can be used to calibrate the deviation setting of an FM signal generator.

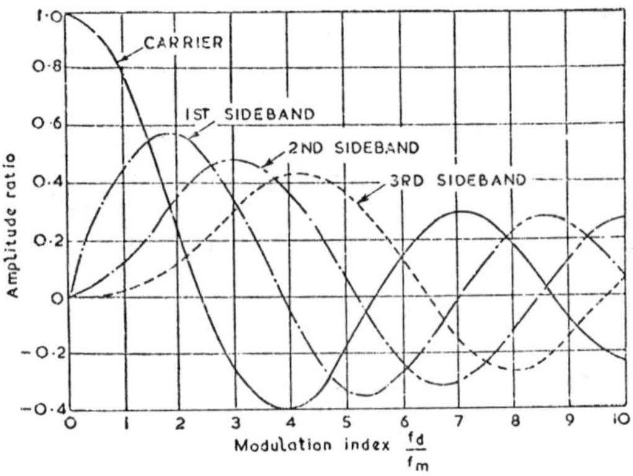

Fig. 8.5 Level of sideband pairs relative to unmodulated carrier

It was mentioned earlier that the simple modulator of Fig. 8.2 that adds a pair of low level sidebands in quadrature to the carrier is adequate for low deviation. The Bessel Function plot shows this to be true, only the first pair of sidebands are significant at low deviation. This corresponds in the phasor diagrams to the fact that a small section of the circumference of a circle is close to a straight line.

5.8.6 Theoretical Analyses of Armstrong modulator

The modulator of Fig. 8.2 can be analysed mathematically as follows:-

$$\text{Multiplier Output} = A \cos \omega_c t . m \sin \omega_m t \quad \text{------} \; 8.1$$

Where $A \cos \omega_c t$ is the carrier and $m \sin \omega_m t$ the modulating signal

$$\text{Output of adder } F(t) = A \cos \omega_c t . m \sin \omega_m t + A \sin \omega_c t \quad \text{----} \; 8.2$$

$$\text{Using trig identity } 2 \cos A \sin B = \sin(B-A) + \sin(B+A) \quad \text{-----} \; 8.3$$

$$F(t) = A \sin \omega_c t + \frac{Am}{2} \sin(\omega_m t - \omega_c t) + \frac{Am}{2} \sin(\omega_m t + \omega_c t) \quad \text{--------} \; 8.4$$

this expression has a similar form to the amplitude modulation expression consisting of a carrier and two sidebands, but how do we know that this expression describes a low deviation angle modulated signal? First it is necessary to derive the expression of an assumed angle modulated signal.

5.8.7 Analyses of an assumed angle modulated signal

For such a signal;

The 'carrier' $f_c = A \cos W_c t$ ----- 8.5

The modulating signal $f_m = m \sin W_m t$ --- 8.6

As the result of phase modulation the instantaneous carrier frequency

$$f_c' = f_c + K m \sin W_m t \quad \text{---} \quad 8.7$$

Where K is a factor that relates the instantaneous modulating voltage 'm' to the frequency deviation fd.

i.e. $f_d = K m$ ------ 8.8

$$\therefore f_c' = f_c + f_d \sin W_m t \quad \text{-----} \quad 8.9$$

The angular deviation θ of the modulated carrier is

$$\int W_c \, dt = 2\pi \int f_c' \, dt \quad \text{-----} \quad 8.10$$

$$\therefore \theta = 2\pi \int (f_c + f_d \sin W_m) dt \quad \text{-----} \quad 8.11$$

$$= 2\pi (f_c t - f_d \cos W_m t) \quad \text{-----}$$

$$\therefore \theta = W_c t - \frac{f_d}{f_m} \cos W_m t \quad \text{------} \quad 8.12$$

Hence the expression for a frequency modulated carrier is

$$f(t) = A \sin \left[W_c t - \frac{f_d}{f_m} \cos W_m t \right] \quad \text{------} \quad 8.13$$

The term $\left(\frac{fd}{fm}\right)$ describes the magnitude of the angular deviation which will be in radians. β is used to represent this term

$$\therefore \quad \beta = \frac{fd}{fm} \quad \text{------} \quad 8.14$$

It is of interest to note that the angular deviation is directly proportional to the frequency deviation and inversely proportional to the modulating frequency.

$$f(t) = A \sin\left[W_c t + \beta \cos W_m t\right] \quad \text{-----} \quad 8.15$$

the plus sign in the above expression is arbitary and changes with the sense of the deviation.

We can now expand this expression to show the sideband structure.

Using the trig. identity

$$\sin(A + B) = \sin A \cos B + \cos A \sin B \quad \text{------} \quad 8.16$$

$$f(t) = A\left[\sin W_c t . \cos(\beta \cos W_m t) + \cos W_c t . \sin(\beta \cos W_m t)\right] \quad \text{-----} \quad 8.17$$

At this point it is wise to leave the full breakdown into sideband terms to a mathematician as the cos of a cos and sin of a cos terms are complicated and lead to the Bessel Function series.

The end result of applying the Bessel Function theory is an expansion into an infinite series of terms.

For the F.M. case the series is:-

$$\begin{aligned} A\Big\{ & J_0(\beta)\cos W_c t \\ & + J_1(\beta)\left[\cos(W_c + W_m)t - \cos(W_c - W_m)t\right] \\ & + J_2(\beta)\left[\cos(W_c + 2W_m)t + \cos(W_c - 2W_m)t\right] \\ & + J_3(\beta)\left[\cos(W_c + 3W_m)t - \cos(W_c - 3W_m)t\right] \\ & + \text{------------}\Big\} \quad \text{----} \quad 8.18 \end{aligned}$$

The values of J_1 J_0 J_2 J_3 etc. can be obtained from the Bessel Function plot of Fig. 8.5 The sum and difference terms represent the pairs of sidebands.

Instead of expanding the relationship using Bessel Functions, it is possible to simplify equation 8.17 for the case of low angular deviation. Certain terms can be simplified as follows for low values of β.

$$f(t) = A\Big[\sin W_c t . \underbrace{\cos(\beta \cos W_m t)}_{\rightarrow 1 \text{ for low values of } \beta} + \cos W_c t . \underbrace{\sin(\beta \cos W_m t)}_{\rightarrow m \cos W_m t \text{ for low values of } \beta}\Big] \text{ -- } 8.17$$

$\therefore$ for low values of β $\quad f(t) = A\big[\sin W_c t + \cos W_c t . \beta \sin W_m t\big]$ --- 8.18

This is an identical expression to that produced for the Armstrong modulator shown in Fig. 8.2, and confirms that for low deviation that arrangement produces an angle modulated signal. The terms that were simplified in equation 8.17 describe the extra side-band terms that would be required for a higher deviation signal, they yield the Bessel Function series for the general case.

5.8.8 Practical Bandwidth Requirements for Angle Modulation

It has been shown that an infinite number of sidebands is necessary to describe an angle modulated signal. In practice a finite bandwidth is acceptable as only a small proportion of the signal power lies in the higher harmonic sidebands.

It is possible to calculate the bandwidth requirements that will encompass a given percentage of the signal power. For convenience a rule is generally applied whereby the bandwidth is set at:-

$2\ fm\ (1+\beta)$ ----- 8.19 this is called the Carson Bandwidth (Bc)

as $\beta = \frac{fd}{fm}$ both the deviation and the modulating frequency are taken into account.

It is worth pointing out that Bc is the total bandwidth requirement to allow for positive and negative deviations of the carrier.

5.9 Phase and Frequency Modulation

Phase and Frequency Modulation are both forms of angle modulation, the difference being the relationship between the angular deviation and the modulating frequency.

5.9.1 Phase Modulation

With Phase Modulation the angular deviation remains constant as the modulating frequency changes. It follows that with such a system the frequency deviation changes as the modulating frequency is varied. Because the angular deviation remains constant the Modulation Index β remains constant.

True phase modulation systems are not normally used for land mobile communication. In general only phase modulators are used with appropriate frequency response shaping networks to convert the characteristic to Frequency Modulation. In such an arrangement it is not possible to maintain a flat frequency response down to d.c. as this would require an infinite phase deviation.

5.9.2 Frequency Modulation

For Frequency Modulation one of the reactive elements in an oscillator must be varied. This is usually achieved with a variable capacitance. Another system uses a reactive feedback path round an amplifier; if the gain of the amplifier is varied, the output impedance has a changing reactance and can be used as one of the reactive elements in an oscillator.

5.9.3 Practical Phase Modulators

It is common practice to phase modulate the output of a crystal oscillator operating at a sub-multiple of the transmitter output frequency. The modulated signal is passed through a chain of multiplier stages to produce the correct frequency. Multiplication factors between 8 and 32 are typical. The higher the multiplication factor the lower the modulating frequency can be for a given frequency deviation.

The phase modulator circuit is usually one of the following.

a) A variable gain amplifier stage with reactive feedback.

b) A variable frequency resonant circuit. If a signal is passed through such a network, it will be subject to a phase shift dependant upon the frequency of the signal relative to the network resonance.

5.10 AUDIO PROCESSING FOR ANGLE MODULATING TRANSMITTERS

5.10.1 Pre-emphasis and De-emphasis

It is common practice with angle modulation systems to employ pre-emphasis of the audio signal in the transmitter and de-emphasis in the receiver. Pre-emphasis and de-emphasis mean respectively an increased or decreased amplification of the higher modulating frequencies at a rate of 6dB per octave.

Pre-emphasis of the transmitted signal is automatically provided if a phase modulator is used. The purpose of this arrangement is to increase the signal to noise ratio at the receiver audio output. The noise spectrum from an FM detector rises at 6dB per octave, if the detector is followed by de-emphasis, the noise spectrum will be made flat. However the de-emphasis will cause the demodulated signal to have a falling response, this is corrected by the rising response at the transmitter. This arrangement gives a significant improvement in signal to noise ratio in the order of 5dB.

5.10.2 Use of a Limiter

Transmitters have to meet stringent specifications laid down by government authorities, these set limits on the signal level allowed to fall into adjacent channels.

A limiter in the path of the modulating signal will set the maximum peak to peak voltage swing; however this will not limit the maximum frequency spectrum produced by the modulator. This is because the limiting action generates higher harmonics which lead to a wider frequency spectrum. A low pass filter attenuating components above 3.3kHz is included after the limiter to remove these higher harmonics due to the limiter.

5.10.3 Pre-emphasis before the limiter

If a speech signal, which consists of both low and high frequency components, is passed through a limiter, the lower level high frequency components will be removed. This is because the high frequency components sit on top of the low frequency waveform. The higher frequency components add considerably to the intelligibility of the speech, and their loss is detrimental.

If the speech signal is pre-emphasised prior to the limiter, then after limiting there will be more zero crossings corresponding to these high frequencies. This will maintain the inteligibility of the speech whilst limiting the maximum deviation from the modulator.

5.10.4 De-emphasis after the Limiter

The pre-emphasis prior to the limiter would mean that at input levels below the limiting level there would be an extra 6dB per octave rise. This can be prevented by including de-emphasis after the limiter. This has the additional advantage of attenuating in-band harmonics produced by the limiter.

5.10.5 Summary of Audio Processing for Angle Modulation Transmitters

The usual arrangement takes the audio signal through pre-emphasis, limiting, de-emphasis and a low pass filter. If a frequency modulator is being used, rather than a phase modulator, then the de-emphasis stage is omitted. This overall pre-emphasis characteristic when combined with the receiver de-emphasis leads to an improvement in the received signal to noise ratio.

This particular sequence of signal processing has been arrived at over the years as a result of both theoretical and practical considerations. Other arrangements have been tried, but many road tests have confirmed that this is the best compromise between inteligibility, complexity and ability to meet the spurious specifications.

5.11 THE COMPARATIVE PERFORMANCE OF THE DIFFERENT MODULATION SYSTEMS

For analogue speech communication in the private mobile radio service all of the modulation methods discussed give an acceptable grade of service. Single sideband systems are not currently used above 30MHz because sufficient radio channels are available and its use will increase the cost of equipment.

The real difference is seen when the radio needs to interface with the public telephone network. Only F.M. with a channel spacing of 25kHz can really meet the standards demanded by this type of service.

Radio wave propogation causes amplitude and phase perturbations on the signal. These changes are never completely removed by the receiver circuits and cause residual distortion components. The subjective effect when listening to these components varies markedly with the modulation methods and reception conditions. With A.M. and SSB the components have appreciable amplitude but their effect is not very disturbing to the listener and users have found these systems quite acceptable. The distortion can remain high even at good received signal levels.

With F.M. the distortion drops to a very low level with an adequate signal, provided the frequency deviation due to modulation is sufficiently high, say 5kHz peak. At low signal levels the distortion caused by noise bursts with F.M. is unpleasant to a listener, however it is by far the most popular modulation technique. The rapid increase in noise at low signal levels can be removed by effective squelch circuits; with such a system the audio signal is turned off when the input signal level falls below a preset threshold.

F.M. receivers and transmitters are generally less complex than those used for A.M. or SSB.

A.M. and SSB are linear modulation systems. This means that the amplitudes of the baseband signals and the corresponding sidebands falling within the receiver passband have the same relative amplitudes. This is significant when the effects of signals on adjacent channels and distant transmitters using the same channel (Co-channel) are considered. Angle modulation is non linear, hence angle modulation receivers tend to reduce the effect of weaker interfer ing signals, thus enhancing the wanted one. This is called the Capture Effect and for narrow band F.M. systems provides a rejection of interferring signals of about 6dB compared with linear receivers. This means that with angle modulation the distance between users of the same channel can be significantly less than that required with linear modulation. This is particularly important in the design of cellular mobile automatic telephone systems.

A special case of interference occurs when large geographical areas have to be covered using the same channel frequency. This is achieved by using a number of transmitters in carefully chosen locations. The individual transmitter frequencies are carefully controlled to produce frequency separations of between 5 and 20Hz. There will always be areas where two or more transmitters arrive at signal strength. With linear modulation the signals add or cancel at a rate equal to the frequency separation of the transmitters. The demodulated signal alters in level but suffers little distortion. The non-linear effect of angle modulation causes considerable distortion in this case. Area coverage systems can be produced using any of the modulation systems described, however angle modulated systems require much tighter control of the modulation parameters and frequencies than linear modulation systems.

Chapter 6

Alternative modulation methods

R.C.V. Macario

6.1 INTRODUCTION

The modulation techniques in use today for land mobile radio (LMR) systems are very practical and well established, especially with regard to voice communication and simple selective calling schemes. Any new or alternative modulation method must therefore clearly show some worthwhile attribute in order that it might, or should be considered. In the single-channel per carrier (SCPC) situation, which is usual for LMR, there are two primary communication resources; the channel bandwidth and the transmitted power. In most situations, the LMR channel is either power-limited or band-limited. Any alternative modulation method must therefore primarily overcome one of these limits at least, i.e., (1) the modulation method occupies a smaller bandwidth for the same communication capability, or (2) offers a better range for the same transmitted power, or a lower interference factor.

The above considerations also apply to group LMR arrangements, such as trunking and cellular schemes. These group technologies at this time are constructed on the basis of a particular SCPC modulation, usually FM. A totally alternative approach is to consider a multi-channel per carrier, or spread spectrum arrangement, where each mobile in a group in effect occupies a much greater bandwidth than an SCPC situation. This technology is considered below in section 6.5.

The primary driving force behind a search or an introduction of an alternative modulation method is nowadays "spectrum efficiency". Periodic reviews of the use of the radio spectrum (1,2,3) always reach the conclusion that private mobile radio (PMR), at least, will always be confined to a restricted amount of the radio spectrum at least below 1000 MHz. Thus spectrum efficiency used to be related primarily to bandwidth, but this view is now changing (4,5). In addition, "efficiency" has several definitions. Spectrum efficiency refers to considerations such as the "number of users receiving satisfactory service per square mile per MHz". Spectrum efficiency is important when many users require access to the spectrum resource, but usually spectrum efficient equipment is more costly than less efficient equipment. Economic efficiency deals with the notion of "highest value used". In the communications field this applies to the market response of the supply and demand for spectrum. Economic efficiency can be encouraged so that equipment appropriate for localized conditions is put into place by individual system operators. Administrative efficiency should also be considered. This refers to the costs to applicants and licensees of meeting requirements for construction permits, renewals etc., and the cost to the Licensing Authority of evaluating these submissions.

Implementation of new technologies (modulation) can also be limited by several factors. Two specific examples are, (1) mandatory technical standards that prohibit the implementation of new systems, and (2) insufficient incentive for users to employ spectrum efficient systems. Some progress has been made in the last several years to eliminate these barriers. Positive steps include:

Allowing some users <u>technical flexibility</u> in the types of systems they may operate.

Use of <u>exclusive channel assignments</u> so that users may benefit from improvements they make.

Encouraging efficient operation by allowing <u>profit-making</u> and resale of excess capacity.

The provision of an exclusive licence, with an associated channel assignment avoids the problem of one type of modulation restricting or interfering with the alternative type (6,7,8).

If one considers that voice communication is the primary objective of LMR, usually confined to a base bandwidth less than or equal to 3 kHz, the matter of voice waveform or signal digitization must be considered. If the voice waveform is left in the analog form, then only a few alternative ANALOG modulation methods are possible, which are considered below. If, on the other hand, the speech signal is digitized by some means or another, DIGITAL modulation methods can be considered. Very many alternative modulation methods can then be described, and an overview of those methods, which are compatible with the LMR scene, is given below. The major step between these two categories of modulation is, of course, that of voice digitization or speech coding as it is sometimes called. A brief review of the possibilities in this area is included because the technique of speech coding affects both the bandwidth requirements of the LMR channel and the acceptab ility of the received voice signal.

6.2 ANALOG MODULATIONS

For a narrow band LMR channel there is very little choice from either conventional AM or low-index FM or PM, as Fig.6.1 would at first sight indicate. The concept of a "reduced" carrier is used in this diagram. This can be a diminished carrier, i.e., 13 dB below PEP, or a "suppressed carrier", i.e., >30 dB below PEP; later a variable reduced level is described.

SSB has been described in the previous chapter and whether a diminished carrier (9), an above-band tone (10), or a tone-in-band (11) system is employed, the system performance, except in certain specific details, do not vary too greatly. Reference to specific Laboratory and field tests on SSB systems has already been mentioned (7,8).

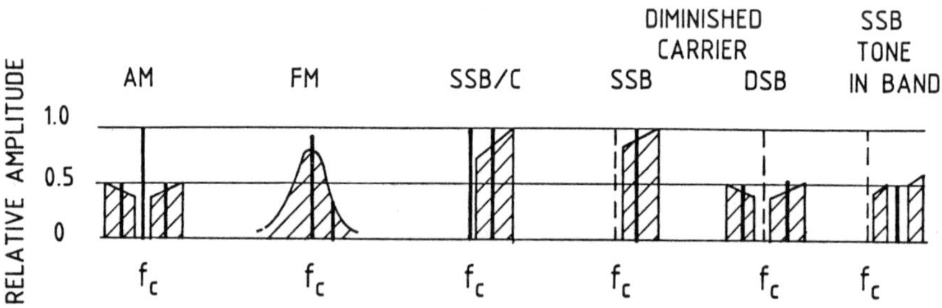

Fig.6.1 Spectra of the various types of modulation

A comparison of the group of analog modulations is shown in Fig.6.2 (12). Narrow band FM (NBFM) initially gains in the intelligibility rating, but the results were based on a heavily processed FM voice signal, which collapses much more quickly as the input carrier-to-noise ratio of the received signal worsens. CPM stands for carrier-position-modulation, otherwise known as phase-locked SSB (PL-SSB). This is a form of SSB which has a full carrier component when there is no voice waveform; when there is a voice waveform the signal becomes SSB-SC. In practice, severe distortion or bandspreading of the signal waveform occurs at the voice signal threshold, so CPM is not a recommended alternative, if one is strictly adhering to a narrow band modulation.

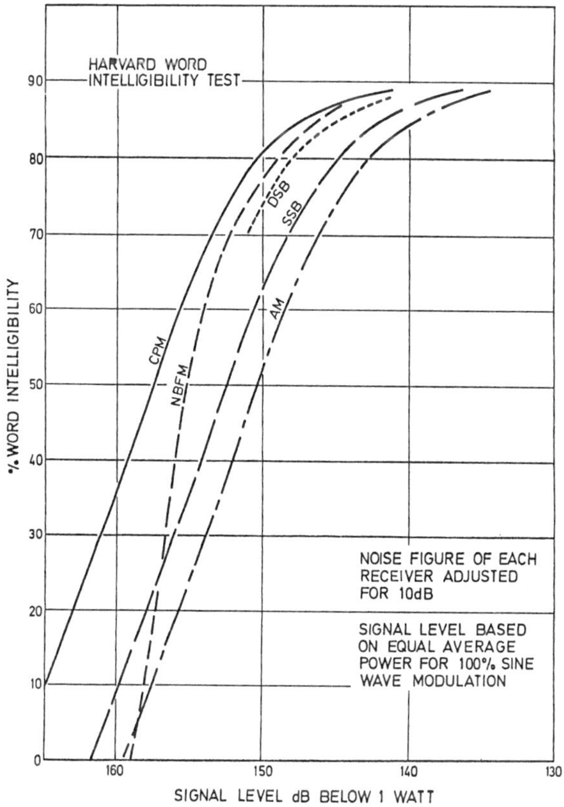

Fig.6.2 Range comparison of equal power alternative analog modulations based on word recognition

Double-sideband suppressed-carrier (DSB-SC) has a 4.8 dB power advantage over conventional AM and a natural carrier lock capability (13). DSB-SC or diminished-carrier (DSB-DC) can have a high intelligibility rating as far as LMR is concerned, but the rapid envelope fluctuation is a difficulty for the receiver AGC and hence a controlled carrier (DSB-CC) form has been proposed (12), as has also been SSB-CC (14).

Fig.6.3 shows the waveform and spectrum of DSB-CC. The signal bandwidth is exactly the same as conventional AM, but has the 4.8 dB power advantage. Fig.6.4 shows the aid DSB-CC offers to a receiver's AGC circuits, in the guise of the average rectified signal level versus depth of modulation. The average power of conventional AM is, of course, constant whatever the modulation.

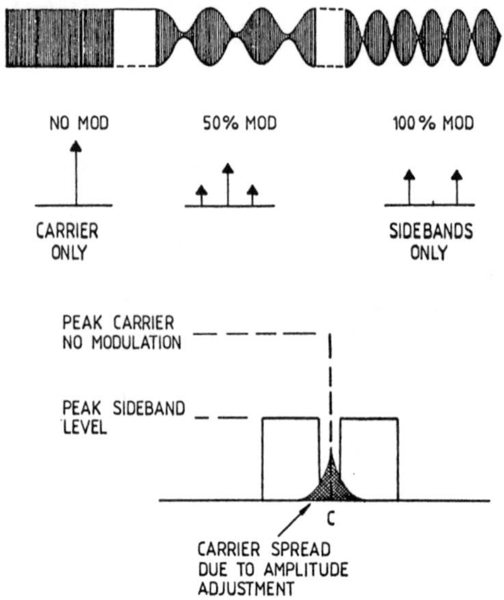

Fig.6.3 Signal Spectrum of DSB - Controlled Carrier

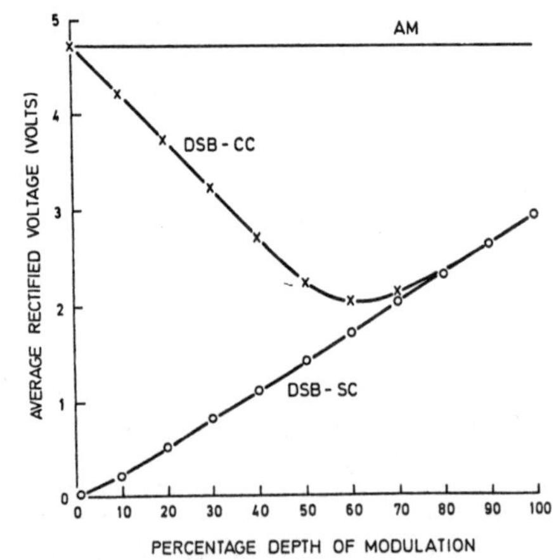

Fig.6.4 Measured Average Rectified Voltage For Three DSB Modulations

A method of generating DSB-CC is shown in Fig.6.5. It would normally require to be followed by a linear transmitter amplifier (15), but a conventional AM mobile with the much more efficient Class C output stage can be converted for DSB operation (16). The equation on which

Figs.6.4 and 6.5 are based is:-

$$s(t) = (1-m)\cos\omega_c t + m\cos\overline{\omega_c \pm \omega_m}.t$$

where m = depth of modulation, 0 › 100%

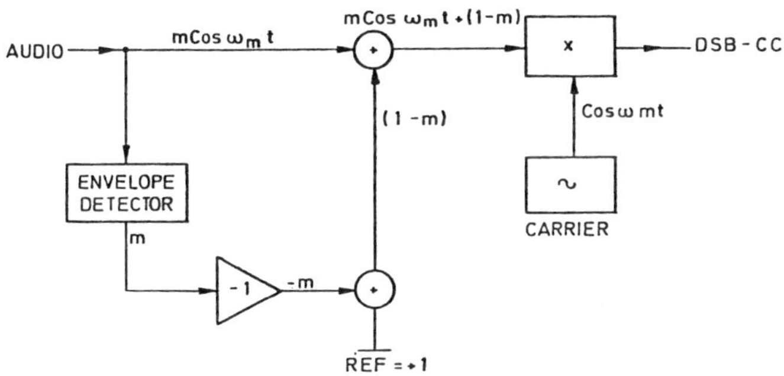

Fig.6.5 A Method of Generating DSB-CC

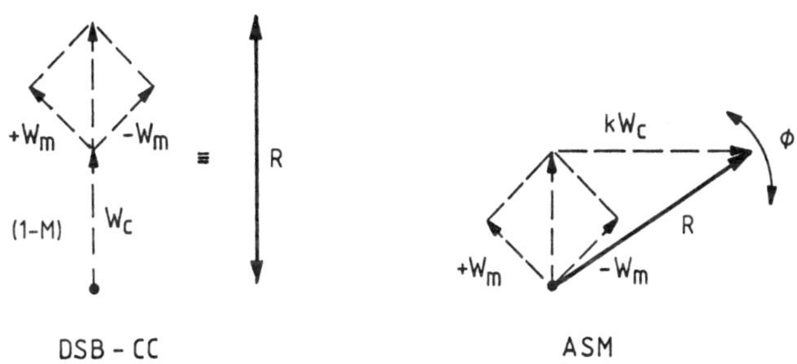

Fig.6.6 Phasor Structure of DSB-CC and ASM

The reason for showing Fig.6.5 is also to introduce a little known analog modulation called arc-sine modulation (ASM), (17). Here the angle of the carrier is modulated by the arc-sine of the modulation. The phasor diagram of ASM is shown in Fig.6.6. ASM is formed by adding a modulated quadrature-phase carrier component to a DSB-SC signal. DSB-CC is formed by adding a modulated in-phase carrier component to a DSB-SC signal, as shown in Fig.6.6.

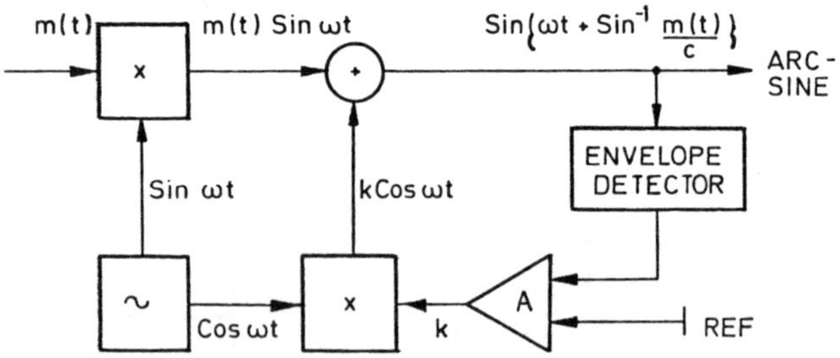

Fig.6.7 Arc-Sine Modulator

A method of generating ASM is shown in Fig.6.7. The similarity to Fig.6.5 is very apparent. ASM has the advantage over conventional FM and PM in not being distorted by amplitude and phase variations in the signal path. ASM is an example of a constant envelope modulation of which several examples are being re-invented (18).

There are also several "pre-distorted" analog modulations such as single-side band FM and compatible single-sideband (CSSB). These modulations aim to reduce the bandwidth requirements of the SBSC signal, but retain an easily implemented non-coherent detector in the receiver. The mobile radio path and inevitable IF phase and amplitude response distortions in a (mobile) receiver react against these modulations, so that their expected benefits are never fully achieved.

A study of the above mentioned analog modulations from the viewpoint of the drive (battery) power required for the different modulations to provide an equal service area also shows that, whereas a particular modulation can perform well on a given effective intrinsic radiated power (EIRP) basis, in practice, the overall efficiency may be far less, as Fig.6.8 illustrates (19).

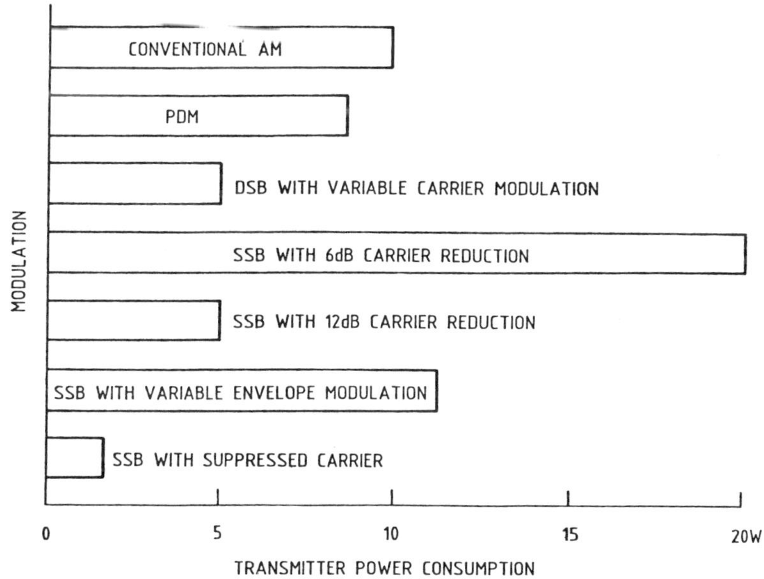

Fig.6.8 Consumption of Transmitters Serving Equal Areas

6.2.1 Demodulation and other considerations

The advantage of conventional AM and FM is that both modulations require relatively simple demodulators in the receiver. Most of the alternative modulations listed above require coherent demodulation. This is a topic about which much could be written, but suffice it to say that the distortion associated with a mis-tuned reception of an SSB signal is very well-known and understood. In the LMR environment, there is enough signal impairment in the propagation environment and the mobile's acoustic environment to dissuade the need for any additional distortion, and therefore an added penalty, i.e. cost, maintenance, etc., for any alternative analog modulation is the need for coherent demodulation.

This process can be divided into two categories; (1) synchronous, or (2) carrier enhancement.

In the former the phase-lock-loop (PLL) and the Costas Loop, shown in Fig.6.10(a) are well known (20). In the second category, Fig.6.10(b) the incoming carrier is acquired and enhanced by either filtering or a carrier recovery loop (21). The advantage of the carrier enhancement scheme is that they tend to have a soft-fail characteristic, whereas with a PLL and the built-in VCO, a drop-out-of-lock situation can occur, which causes heterodyning in the receiver.

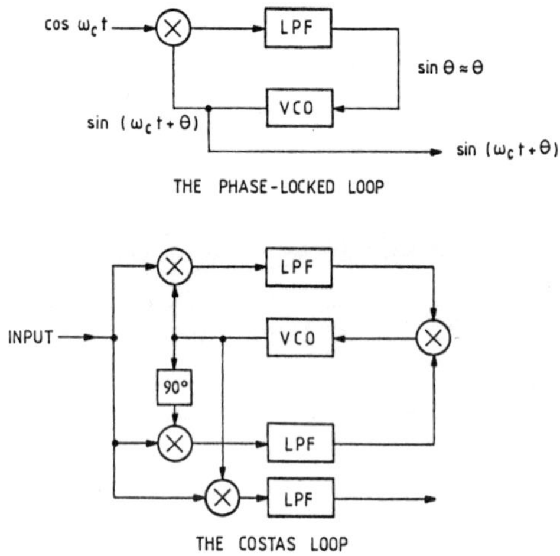

Fig.6.10(a) Synchronous Demodulators

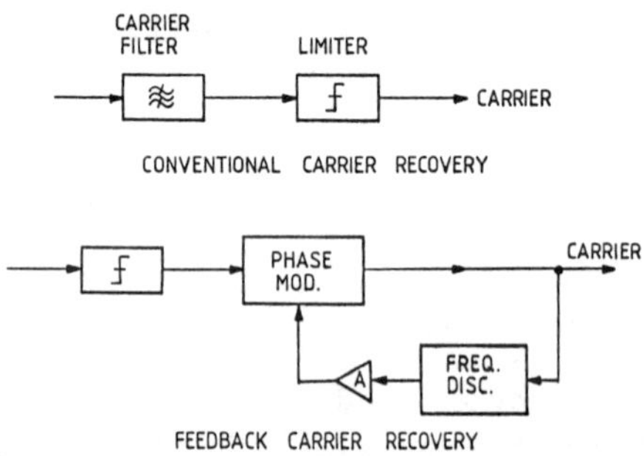

Fig.6.10(b) Carrier Enhancement Demodulators

6.3 SPEECH ENCODING

Before one can venture into digital transmission schemes of voice communication one needs to have some appreciation of the methods and attributes of speech waveform digitization or encoding, as it is know. For example, it is well known that any bandlimited waveform can be represented as a set of impulses, provided the number of impulses per second is sufficient. For speech limited to a 3 kHz upper frequency this would imply 6000 impulses per second. However the impulses have also to be height coded i.e. pulse code modulated (PCM) and clearly the number of pulses per second required in principle can become prohibitively high i.e. for an 8-bit code, this implies 48 kb/s. Fortunately however speech is a fairly redundant form of message transfer and for communication purposes a smaller number of impulses can be accepted i.e. 16 kb/s. Schemes of this type are known as WAVEFORM CODING. For PMR the two most acceptable types are adaptive delta modulation (ADM) and adaptive differential PCM (ADPCM). However if the bit rate is lowered too far the perceptual quality of the speech quickly deteriorates e.g. Figs.6.11 and 6.12 (22, 23, 24). For PMR the easy-to-implement waveform coders are now in use down to about 12 kb/s.

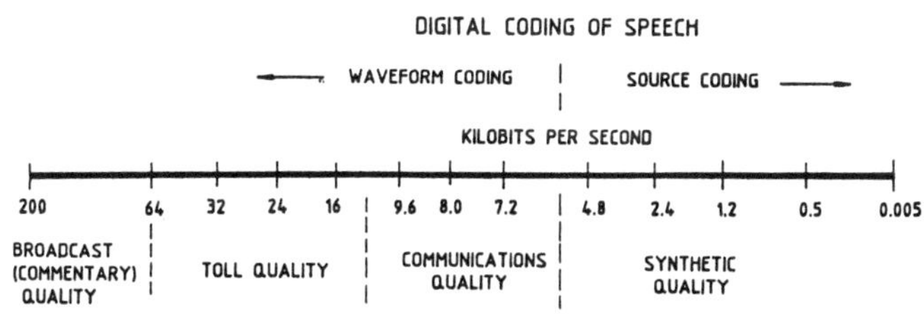

Fig.6.11 Spectrum of Speech Coding Transmission Rates (Nonlinear Scale) and Associated Quality

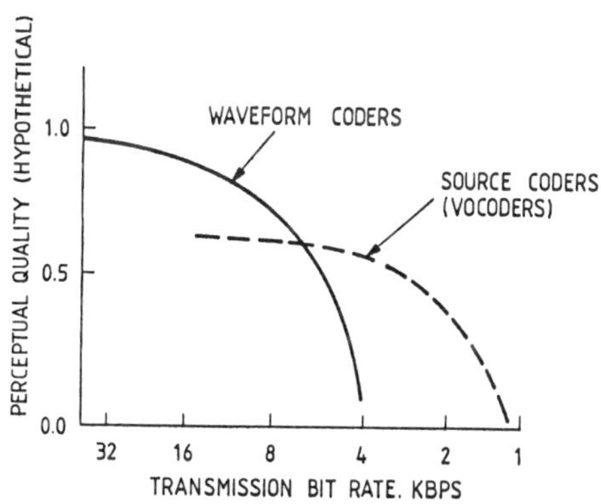

Fig.6.12 Speaker Recognizability

The alternative approach to encoding a speech waveform is to more closely re-construct the particular attributes of the nature of the human voice waveform i.e., at source; hence the name SOURCE CODING. As Figs.6.11 and 6.12 suggest the voice waveform is now partially synthesized and tends to take on a synthetic quality, because the constituent parts in different frequency bands are to some extent computer generated from the sparse (low bit rate) information received. Bit rates usually around 2 kb/s are feasible, but as has been said all vocoders sound like vocoders (22), as well as involving a heavy hardware penalty.

As a compromise MIXED CODERS are being mainly investigated today. Here the best aspects of the above two coders are combined to produce a lower bit rate signal with a higher quality than a waveform coder on its own. The best candidate in this area is known as adaptive or linear predictive coders (APC or LPC), and although more complex than ADM, LPC can produce a higher perceptual quality speech signal than ADM or ADPCM at the same medium bit rate, e.g. 8 kb/s.

There are as yet no CCIR standards for speech coders at these low data rates and therefore application to PMR will tend to be confined to special or special need systems. To illustrate this point it should also be pointed out that the more complex coders are sensitive to voice type and for acceptable operation the speaker should have a 'good LPC voice' for example (24).

6.4 DIGITAL MODULATION

Having organized the communicator's voice into a digital stream of ones and noughts (if a two level code is used) some form of digital modulation of the RF carrier is now required.

There are very many forms of digital modulation. The Table of Fig.6.13 attempts to summarize the situation. There is now a fairly well agreed standard of designation, which is set out in this Table, i.e. references 25 and 26.

TYPE	DESIGNATION	FULL NAME
AM	OOK	ON-OFF KEYING (0,1)
	(PSK)	SUPPRESSED CARRIER AM (+1,-1)
	QAM	QUADRADURE AM (TWO PSK SIGNALS)
	QPR	QUADRADURE PARTIAL RESPONSE (+1,0,-1)
FM	FSK	FREQUENCY SHIFT KEYING
	CP-FSK	CONTINUOUS PHASE FSK
	MSK	MINIMUM SHIFT KEYING = CP-FSK (β=0.5)
	TFM	TAMED FM = MSK+FILTERING
	GMSK	GUASSIAN FILTERED MSK
PM	PSK	PHASE SHIFT KEYING
	DE-PSK	DIFFERENTIALLY ENCODED PSK
	DPSK	DIFFERENTIAL PSK
	QPSK	QUARTERNARY PSK
	OK-QPSK	OFF-SET KEYED QPSK (sometimes called SQPSK and similar to QAM)
	IJF-QPSK	INTERSYMBOL INTERFERENCE and JITTER FREE QPSK

Fig.6.13 The names and designations of the more common shift-keyed modulations

A primary objective of a spectrally efficient modulation is to maximize the bandwidth efficiency, which is defined as the data rate to channel bandwidth - bps/Hz.

The current aim is a transmitted bit rate almost equal to half the value of the SCPC bandwidth, i.e., a 12 kb/s digital system can be accepted within a 25 kHz channel allocation.

A second objective is for the waveform to have a constant envelope. This search for constant-envelope, bandwidth-efficient, digital-modulation scheme has narrowed down the contenders and therefore we will concentrate on the most likely candidates.

Examination of the table of Fig.6.13 will have also indicated that although the various forms of modulation have been grouped in the conventional AM, FM, PM categories, which is correct, the individual modulation types often have a great similarity, e.g., PSK, because the modulation structure, unlike that for analog modulation, has only two, or sometimes three, states namely +1, 0 and -1. The reason for the different modulation types in fact often comes about because of the need for coherent detection, as opposed to non-coherent detection at the receiver. With the advent of modern high speed VLSI circuitry the demarcation between the two schemes becomes less important from a user's point of view, but none the less the detection and bit recovery procedure for most digital modulation schemes is complicated, and beyond the scope of the present text.

For SCPC LMR, one is looking for, as already stated, a constant-envelope bandwidth-efficient high-speed modulation. Both PSK and FSK have a constant envelope. Fig.6.14 shows the basic PSK and FSK waveform over one bit intervals.

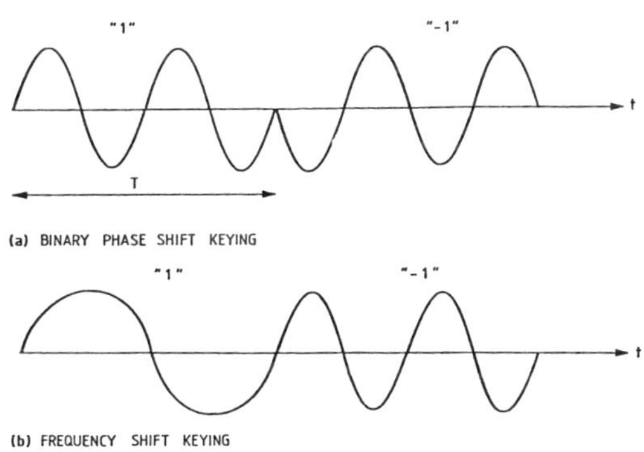

Fig.6.14 (a) Binary Phase Shift Keying;
(b) Frequency Shift Keying

The problem is to merge the two opposite signals (states) smoothly together so that there is not an abrupt waveform transition at the data change which will cause severe baseband spectrum spreading, with the result that the digital modulated waveform is no longer bandwidth efficient.

To illustrate this process we will go quickly through the development of the more efficient modulations (27). Thus in Fig.6.15 is shown the basis of quaternary phase-shift keying (QPSK) which is two PSK signals running in parallel, 90° in phase, which is a perfectly acceptable mode of operation provided coherent demodulation is arranged. The problem with QPSK is the irregular phase transitions, i.e., at 2T, 4T, etc. If the input data streams are off-set or staggered, this leads to OKQPSK which has more orderly transitions.

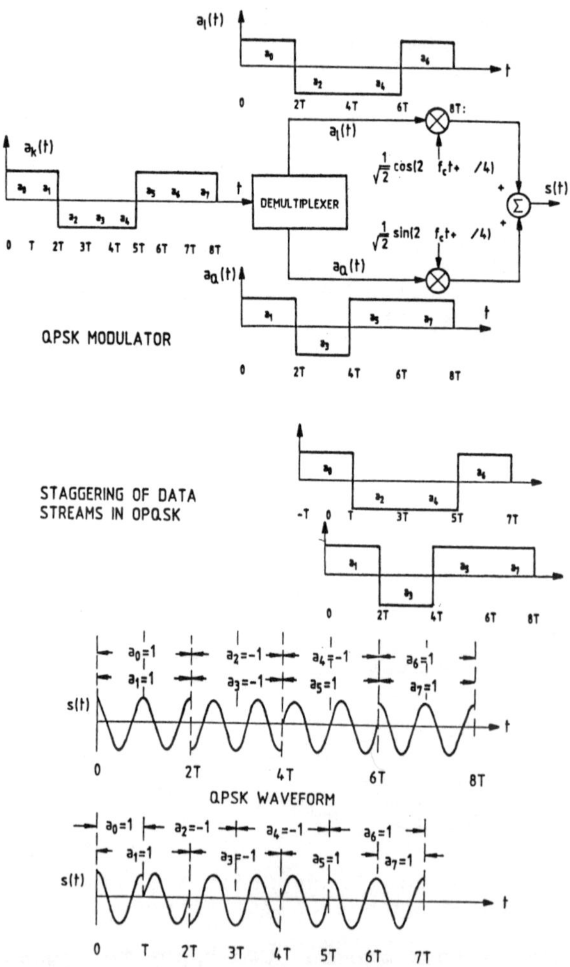

Fig.6.15 The generation and waveforms of QPSK and OK-QPSK

If the data bits, in addition to being off-set, are also pre-shaped as half cosine pulses, this leads to what is known as minimum shift keying (MSK). This is equivalent to FSK with a frequency deviation equal to half the bit rate. The waveform of MSK is shown in Fig.6.16.

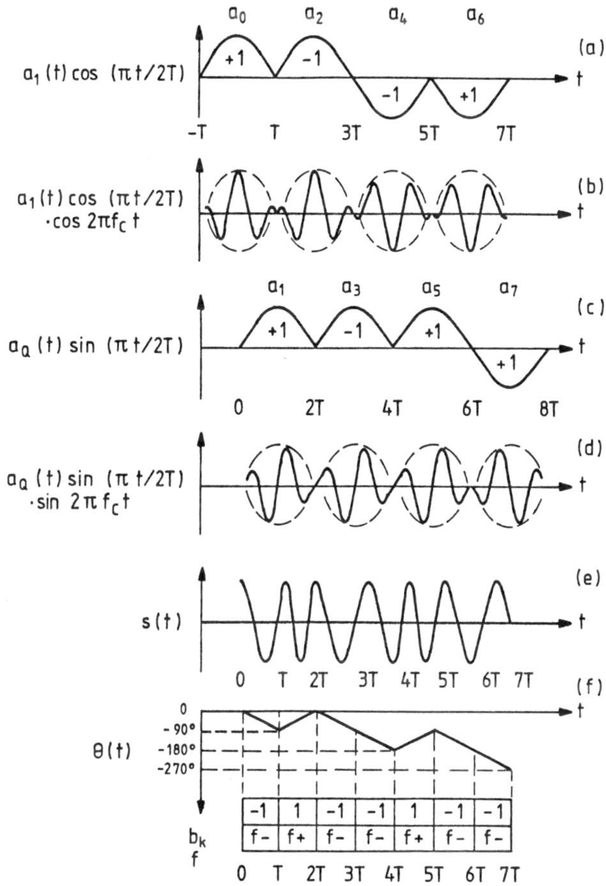

Fig.6.16 The structuring of the minimum shift-keyed FSK waveform

Note the smooth transition from one data bit to the next. This has a marked effect on the power spectra of the transmitted waveform, as Fig.6.17 illustrates. The vertical scale is the output spectral density; the horizontal scale is the normalised frequency difference from the carrier centre frequency, and T corresponds to the data signalling rate 1/T bauds. Even with the improvements of MSK over QPSK, this diagram shows that the severe requirements with respect to out-of-band radiation for a SCPC mobile radio could not be met.

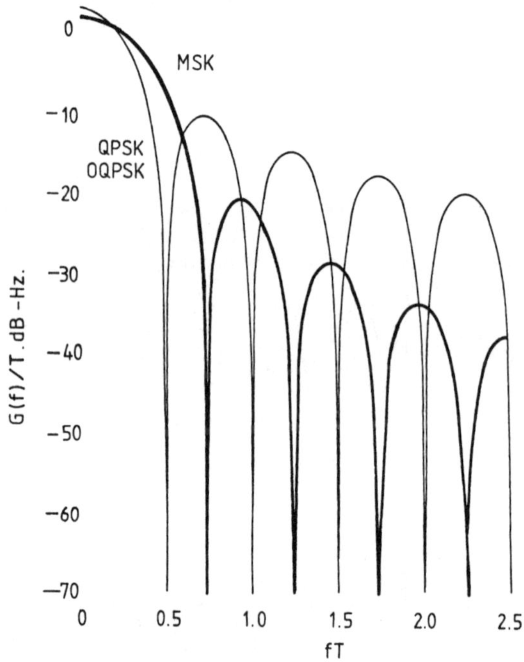

Fig.6.17 The power spectral density against carrier frequency off-set for three digital modulations

Further pre-signal filtering must be applied. Two current methods are tamed FM (TFM) (28), or Gaussian filtered MSK (GMSK) (29). The pre-conditioning filter can be adjusted to give various degrees of adjacent channel performance, as Figs.6.18 and 6.19 illustrate. The systems actually sacrifice the intersymbol interference margin against out-of-band spectral density; however it is clear that conventional adjacent channel interference figures are obtainable with these types of modulations, for example, a bit rate of 12 kbps can be maintained in a 25 kHz channel spacing.

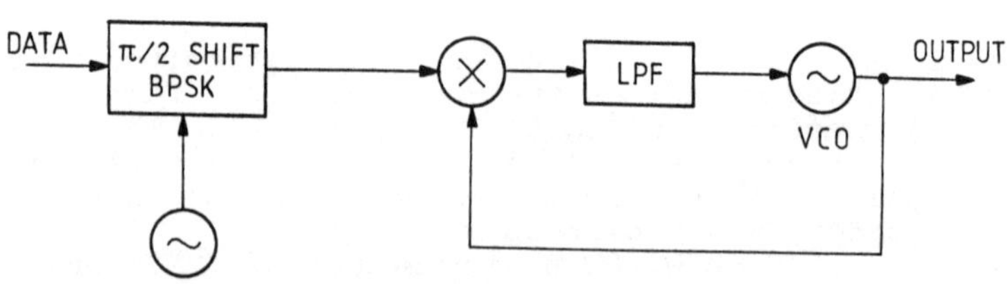

Fig.6.18 PLL-Type GMSK Modulation

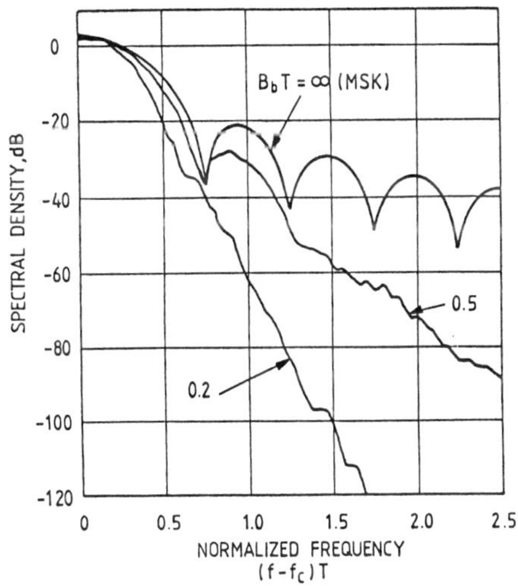

Fig.6.19 Power Spectra of GMSK

These newer 'tamed' digital modulations can approach a transmitted bit rate of nearly one half the SCPC bandwidth, and also not exceed the adjacent channel interference criteria. Table, Fig.6.20 attempts to summarize the present situation (25). This one has the main modulation modes repeated plus three columns. The first column gives the published (sometimes measured, sometimes calculated) received carrier-to-receiver noise ratio for a bit error rate (BER) of 10^{-4}, i.e., one symbol in 10,000 in error. It is noticeable, for example, that TFM as compared to MSK actually sacrifices service range for adjacent channel performance.

Column two gives an indication of the relative bit rate per channel Hz, rounded to a nearest unit. For example, QAM, MSK and QPSK all transmit at twice the basic rate. However, as column three indicates, more speed and less adjacent channel spectral density can mean a higher required carrier-to-noise ratio under fading (mobile) conditions. The BER in this column is now one bit wrong in every hundred bits on average, which is quite disruptive on coded speech. Overall, however, the digital modulation outlined presents an attractive alternative for the LMR scene, and to some extent, their introduction mainly hinges on an acceptable process of speech coding as discussed above. For example a detailed account of a digitized speech transmission experiment using existing 25 kHz channel VHF FM mobiles has been described by Petrovic (30). Here a 12 kb/s ADM speech encoder was used in conjunction with duobinary FSK (31) which met the criteria of having a bit rate equal to one half the channel bandwidth as well as good adjacent channel performance, though the experiment did show that the adjacent channel interference effects were to some extent worse than conventional FM. The newer

TYPE	MODULATION SCHEME	(1) C/N @ BER 10^{-4}	(2) SPEED b/s/Hz	(3) FADING Avg C/N dB
AM	OOK-ENVELOPE DETECTION	12.5	1	17
	QAM	9.5	2	14
	QPR	11.5	2.5	
FM	FSK-NONCOHERENT DETECTION (d=1)	12	1	20
	CP.FSK-NONCOHERENT DETECTION (d=.7)	11	1	18
	MSK (d=.5)	9.5	2	14
	MSK-DIFFERENTIAL ENCODING (d=.5)	10.5	2	17
	TFM	15	2	
	GMSK	11.5	2	18
PM	BPSK-COHERENT DETECTION	9.5	1	14
	DPSK	10.5	1	17
	QPSK	10	2	13
	8.ary PSK-COHERENT DETECTION	13	3	17

(1) Rounded to half dB
(2) Rounded to nearest unit
(3) For BER = 10^{-2}

Fig.6.20 Table comparing the service range of several digital modulations based on published carrier-to-noise data

'tamed' modulations (28,29) are likely to be better in this respect. On the other hand the reported information does prove the point that digital LMR systems are unlikely to dislodge conventional 12½ kHz schemes unless a satisfactory solution to the mixed or source voice encoding technology can be realized in a commercially acceptable manner.

6.5 SPREAD SPECTRUM SYSTEMS

So far this chapter has only considered that every land mobile radio system works on the basis of transmitting the voice message, or digital message, or selective calling code within a single allocated radio channel which is either less or equal to 25 kHz band width, i.e., the SCPC situation. For a large number of users several channels will be allocated on both a time and area basis. These arrangements lead to either trunking (dynamic channel assignment - DCA) or cellular schemes, which are described in detail in following chapters. The alternative modulations described above are applicable to these schemes; for example, the use of SSB in a cellular scheme has been described in detail (32), but usually 12/25 kHz FM is specified, although an all-digital modulation system would be preferable from the point of view of organizing the system protocols.

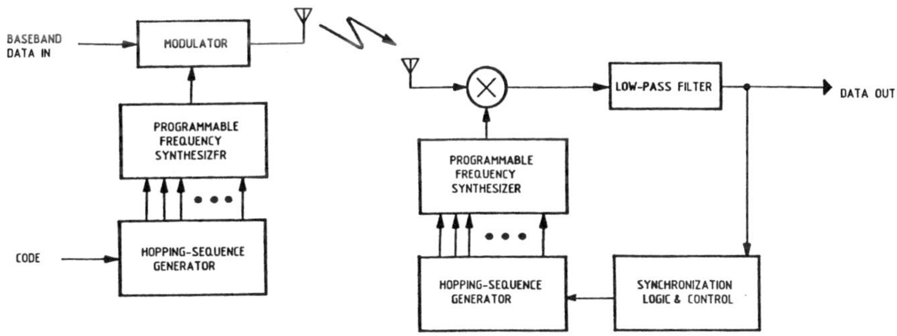

Fig.6.21 FDMA spread spectrum LMR system

An entirely alternative approach is to allocate the entire band allocation (N-channels) to each user, so that he broadcasts, and receives, a very wide band (spread spectrum) signal - almost on par with VHF FM broadcasting. To distinguish one user from another however each signal is partitioned with a unique time/frequency code overlay modulation.

The simplest scheme to appreciate is a frequency-division multiple-access (FDMA) system, though this is not the first spread spectrum LMR system to be proposed (33). Fig.6.21 illustrates the principle.

A conventional mobile is organized to hop from channel to channel in the allocated band on a pre-determined basis - the mobile's code (34, 35) - in a specified time frame. A second mobile could operate in parallel since codes can be chosen such that the likelihood of the two mobiles occupying the same frequency (channel) slot at the same time would be very small: however there are several practical limitations to the ideal theoretical model.

A second approach is to overlay the mobile's message signal on a very high-speed (wide band) bit stream which contains the mobile's code, i.e. Fig.6.22.

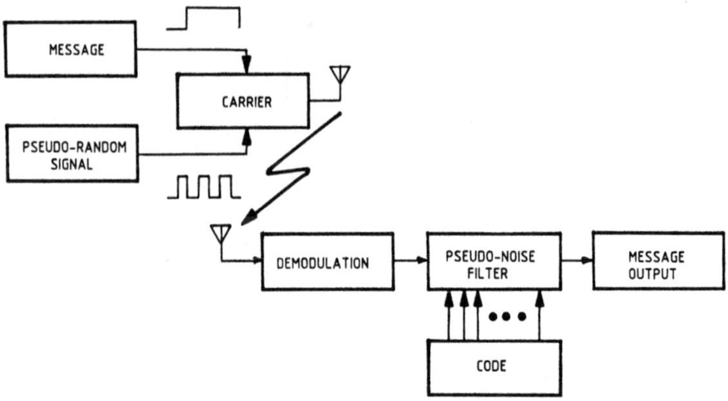

Fig.6.22 TDMA spread spectrum LMR system using PN codes

Usually a pseudo-noise (PN) code is used as the master bit stream (36) and the result is a time-division multiple-access (TDMA) system. This type of system has been studied for military communications because the signal appears as noise to an un-licensed listener but it does have the problem of requiring power control of the transmitters because of what has been termed the near-far effect.

An in-between approach to the concept of spread spectrum LMR is code-division multiple-access (CDMA). Here frequency hopping and channel coding could be combined. For example a FH/CDMA system avoids the problem of trying to settle a mobile on a narrow specific frequency for a short time, i.e. Fig.6.21. Each system 'novelty' however has a cost penalty which is well described by Yue (34). The likelihood of these systems being introduced in the general LMR market from a simple spectrum efficiency view point is very small, but such systems could emerge where there were special requirements, such as an overlay system where the propagation conditions, i.e. VHF Band I, render normal channel planning procedures inappropriate.

REFERENCES

(There are very many references on "alternative modulation": the ones listed are those which in general will enable the reader to search further with the greatest ease).

1. "The use of the radio spectrum", 1977, IEE Report (available from the IEE).

2. "Terrestrial land mobile services", 1982, Report of the Mobile Radio Committee, HMSO publication.

3. "Independent review of the radio spectrum" (30-960 MHz), 1983, The Merriman Review, HMSO publication No. 9000.

4. Berry, L.A., 1978, "Measuring spectrum use", Wireless World, 43.

5. Tou, C.P, and Roy, D.A., 1980, "On efficient spectrum utilization from the standpoint of communication theory", Proc. IEEE, 68, 1460.

6. "Implementing new technology in the land mobile radio service", 1983, FCC/OPP Report, 83-5.

7. "Amplitude companded SSB compared to conventional FM for VHF mobile radio", 1983, FCC/OST Report TH83-7.

8. "Report on comparitive tests of modulation methods for PMR", 1982, Home Office (U.K) Directorate of Radio Technology Report.

9. Wells, R., 1978, "SSB for VHF mobile radio at 5kHz channel spacing", Proc. IERE Conf. Radio Receivers, Southampton, U.K., 29.

10. Lusignan, B.B., 1978, "SSB for land mobile radio", IEEE Spectrum, July, 33.

11. McGeehan, J.P., and Bateman, A.J., 1984, "Phase-locked transparent tone-in-band SSB for mobile radio", IEEE Trans. COM-32, 81.

12. Macario, R.C.V., 1973, "Unusual forms of analog modulation", Wireless World, 79, 281.

13. Macario, R.C.V., 1973, "A VHF receiver adapted for the reception of DSB-SC transmissions", The Rad. & Elec. Eng., 43, 407.

14. Barnes, G.W., 1954, "A single-side band controlled carrier system for aircraft communications", Proc. IEE, 101, 121.

15. Krauss, H., et al., 1980, "Solid state radio engineering", Wiley & Sons.

16. Petrovic, V., and Gosling, W., 1974, "Design of high-efficiency VHF DSB-DC transmitters having low spurious emission", Proc. IEE, 120, 95.

17. Schemel, R.E., 1966, "Arc-sine modulation - a method of overcoming non-linearities in transmission systems", Elec. Letts., 2, 10.

18. Boutin, N., 1984, "Constant envelope signal generator with improved phase linearity", IEEE Trans. COM-23, 488.

19. Petke, G., 1982, "Economic modulation processes for AM broadcast transmitters", EBU Review, 193, 114.

20. Spilker, J.J., 1977, "Digital communications by satellite", Prentice-Hall Inc., 312.

21. Craine, J.F., and Macario, R.C.V., 1973, "Receiver innovations for single-side band broadcasting", The Rad. & Elec. Eng., 43, 312.

22. Flanagan, J.L., et al., 1979, IEEE Trans. Comms, 27, 710.

23. Holmes, J.N., 1982, "A survey of methods for digitally encoding speech signals", The Rad. & Elec. Eng., 52, 267.

24. Bylanski, P., and Chong, T.W., 1984, "Advances in speech coding for communications", GEC Journ. Research, 2, No. 1, 16.

25. Oetting, J.D., 1979, "A comparison of modulation techniques for digital radio", IEEE Trans. Comms., 27, 1752.

26. Stremler, F.G., 1982, "Introduction to communication systems", Addison-Wesley 2nd Ed., Ch. 10.

27. Pasupathy, S., 1979, "Minimum shift keying: a spectrally efficient modulation", IEEE Comms. Magz., July, 14.

28. Chung, K.S., and Zegers, L.E., 1983, "Generalized tamed frequency modulation", Philips Telecom Review, 41, 82.

29. Murota, K., and Hirade, K., 1981, "GMSK modulation for digital mobile radio telephones", IEEE Trans. Comms., 29, 1044.

30. Petrovic, P.M., 1982, "Digitized speech transmissions at VHF using existing FM mobile radios", IEEE Trans. Veh. Tech., 31, 76.

31. Lender, A., 1963, "The duobinary technique for high-speed data transmission", IEEE Trans. Comms. Elec., 82, 214.

32. Prabhu, V.K., and Steele, 1982, "Frequency-hopped single-sideband modulation for mobile radio", B.S.T.J., 61, 1389.

33. Dixon, R.C., 1976, "Spread spectrum systems", Wiley.

34. Yue, O-C, 1983, "Spread spectrum mobile radio, 1977-1982", IEEE Trans. Veh. Tech., 32, 98.

35. Cooper, G.R., and Nettleton, R.W., 1983, "Cellular mobile technology: the great multiplier", IEEE Spectrum, June, 30.

36. Golomb, S.W., 1964, "Digital communications with space applications", Prentice Hall.

Chapter 7

Area coverage techniques

R.J. Holbeche

7.1. Introduction.

Historically the vast majority of mobile radio schemes have developed from a single base station and a small number of portables. However, certain types of users have always had a requirement to operate over a geographic area larger than could in any way sensibly be covered by a single base station. This requirement is generated primarily in the public sector by the police, fire and ambulance authorities although a number of other major user groups including the utility organisations have developed an increasing reliance on radio, now producing a requirement for multi-transmitter area coverage schemes. The need for such area coverage schemes is especially important at UHF where many portable equipments are used.

Clearly the way in which these multiple transmitters are configured and operated will totally define the characteristics and performance of the resulting area coverage scheme.

7.2. Multichannel Schemes.

By far the easiest and most straight-forward way of achieving area coverage is to install a sufficient number of base stations so that every required part of the service area is covered by at least one of them. Each base station transmitter is then operated on a different channel, and are then operated simultaneously with the same modulation.

All that is required of the mobiles is that they switch manually to the appropriate channel depending on their geographic location. However, the vast majority of mobile radio users are not trained radio operators and are frequently more concerned with performing their proper job function than with operating the radio. Also, the individual coverage areas of each transmitter are most likely quite different from other more familiar geographic sub-divisions. Overall the scope for lost calls and unnecessarily poor communications is considerable.

Multi-channel schemes are clearly wasteful of spectrum although they do have the merit that co-channel interference, caused perhaps by anomalous propagation, will tend only to hamper communication in one part of the total service area.

7.3. Basic Co-channel Schemes.

In order to reduce the spectrum requirements and to avoid the need for mobiles to repeatedly switch channel it is possible to operate all base transmitters on the same channel, and so as to avoid any interference effects between them in overlap areas, it is then necessary to operate transmitters individually rather than simultaneously.

Although apparently very simple, the problem arises that the fixed operator will not be aware of the location of the mobiles, so what amounts to a very primitive vehicle location scheme must be incorporated into the system.

This requires both selective calling facilities and a return path voting system (described later) to be incorporated into the system.

The system operates for a call initiated by the fixed station as follows:

(1) Each base station in turn will transmit the appropriate selective call sequence for the required mobile.

(2) The mobile, on receipt of its call sequence, will then transmit normally the same sequence and incidentally alert the driver that a call is imminent.

(3) Each base station receiver that is in range of the mobile will then inform the central control point of the system of the strength or quality of the received signal so that the optimum base station site can be selected for transmission to the mobile.

For calls initiated by the mobile the driver will normally transmit his selective call (equivalent to his callsign) and the system will select the optimum base station as before.

This technique, which is favoured by the message handling organisations who need a selective calling system anyway, provides largely satisfactory performance. In general terms it has complexity as its major disadvantage and its successful operation relies to a large extent on the considerable skill of the fixed operators.

7.4. Spaced Carrier Schemes.

An early attempt at co-channel operation specifically for U.K. police operation used amplitude modulated transmitters with the carriers spaced so that carrier beats in the mobile receiver in overlap areas were at a higher frequency than the upper limit of the audio response to avoid irritating and distracting whistles.

For a three state scheme the carriers were offset -9,-3 and +9kHz from the channel centre frequency. As the total bandwidth required is approximately 24kHz then these schemes had to be operated on 50kHz channelled systems.

Clearly in the overlap areas, although carrier beats were largely eliminated, intermodulation between the different modulations often produced quite severe colouration of the speech. However, this disadvantage was outweighed by the considerable convenience of single channel operation. These spaced carrier schemes have been used for many years by the Public Safety organisations.

The increasingly significant objection to these schemes is their excessive bandwidth requirements.

7.5. Quasi-Synchronous Schemes.

In order to preserve the convenience of co-channel operation but at the same time reduce excessive spectral requirement, it is possible to operate the transmitters with their carrier frequencies offset by much less than the 300Hz lower audio limit of the mobile receiver. These quasi-synchronous schemes can then operate with either amplitude or frequency modulated equipments, and became feasible with the availability of high performance crystal oscillator units.

7.5.1. Quasi-Synchronous F.M. Schemes.

Initially Q.S.F.M. (known in the U.S.A. as Simulcast) schemes were set up with carrier frequency offsets of a few tens of Hertz, representing the performance limitation of quartz crystal oscillator units of the day. It was assumed that due to the F.M. capture effect the mobile receiver would respond only to the strongest signal present and that periods of apparently equal signal strength, when capture would not occur, would be infrequent and produce few problems. In practice it was quickly found that performance was disappointing but could be improved by very carefully matching the radiated modulation with respect to both amplitude and phase. Also, tailoring the coverage of the transmitters so as to avoid even approximately equi-signals in areas where intelligible communication was required, was extremely beneficial.

The difficulty in establishing a successful Q.S.F.M. are considerable. It is marginally easier at U.H.F. than V.H.F. and the overall quality of reception is improved significantly if the frequency offsets are reduced from the few tens of Hertz of the prototype schemes down to a few Hertz of the more recent schemes.

The fundamental problem is that an F.M. receiver produces a high level of audio noise when the signal level falls below the receiver threshold. Although this is of course experienced naturally for single station working during deep fades in weak signal areas, in a Q.S. scheme the repetitive nature of these noise bursts is extremely distracting. The additional and equally significant problem is that the F.M. capture effect is far less pronounced with low-deviation narrowband mobile radio schemes than with high-deviation broadcast systems and quite inadequate to achieve satisfactory performance for Q.S.F.M. schemes without extremely careful modulation matching.

7.5.2. Quasi-Synchronous A.M. Schemes.

A highly significant difference between A.M. and F.M. receivers is that whereas an F.M. receiver will produce a high level of audio noise for a very low input signal level, under the same conditions an A.M. receiver will produce a very low level of audio noise.

Also, whereas an F.M. receiver relies on hard limiting to cope with dramatic variation of input signal level, an A.M. receiver must be equipped with an Automatic Gain Control (A.G.C.) system in order to preserve the correct modulation.

For an F.M. system the modulation parameter (phase) and the signal level (amplitude) are quite different. However for an A.M. system both the modulation and the signal level are the same parameter. Therefore under conditions when the mobile receiver experiences signal components from more than one transmitter, modulation will be additive in the A.M. case whereas any analysis of the F.M. situation is hideously complex.

Q.S.A.M. schemes have been found to be most successful when the frequency offsets are a few Hertz. This is because the A.G.C. system in an A.M. receiver is designed to compensate only for mean signal level variations which in practice occur only slowly. It is generally considered undesirable to increase the speed of response of the system as this would have a detrimental effect on any sub-audio signalling systems.

Careful modulation matching is employed in these schemes, as in F.M. schemes, but the required accuracy of matching is found to be much less. Deterioration of performance occurs more gradually with mismatch in A.M. schemes than F.M. ones, significantly easing the task of installation.

An additional important facility offered by Q.S.A.M. is that of employing "fill-in" base stations to improve coverage in otherwise poor areas. This is possible and perfectly straightforward because careful tailoring of service areas from individual sites is not necessary for A.M. schemes.

The effect in a weak two-signal overlap area in a Q.S.A.M. scheme is a periodic increase in background noise level at a rate equal to the offset frequency. This has only a minor detrimental effect on speech transmissions but is potentially a more serious cause for concern on continuous high-speed data transmissions. A successful technique devised specifically to overcome this effect is called Sideband Diversity. In this system, instead of the modulation being identical at each transmitter, audio phase shift networks are inserted in the audio paths, the ultimate effect of which is to prevent total signal loss in the mobile due to the presence of equal signals. Instead each sideband and the carrier collapse at different times so that an appropriate demodulator process can continuously produce the correct output. Early work used an independent sideband receiver with a simple decision circuit that routed the demodulated larger amplitude sideband to the receiver output. More recent work has involved the development of synchronous detection methods to provide an uninterrupted audio output.

7.6. Synchronous Area Coverage.

The philosophy behind this recent development in area coverage technology involves a consideration of propagation to a mobile or portable in a built-up area where the received signal consists of a number of reflected components. Under these conditions it should make very little difference whether all the received components originate from the same transmitter or from more than one transmitter provided the radiated transmissions are identical in every way. This overcomes the problems of frequency stability at UHF and distortion due to mismatched modulation.

In order to implement a truly synchronous system, equipments are required that can reradiate identical transmissions from a number of remote sites. This has been achieved by using radio links between the control point and the remote sites which are provided with a synchronised non-demodulating repeater, Fig 1. In those equipments the modulation is not recovered from the incoming link signal, but down-converted to an intermediate frequency to provide selectivity and then up-converted to the final radiated frequency. In addition, a facility is provided to lock the equipments master oscillator to the incoming signal.

Synchronous schemes of this type are much easier to install than conventional link controlled Q.S.F.M. schemes and provide at least as good a quality of overall performance as Q.S.A.M. schemes.

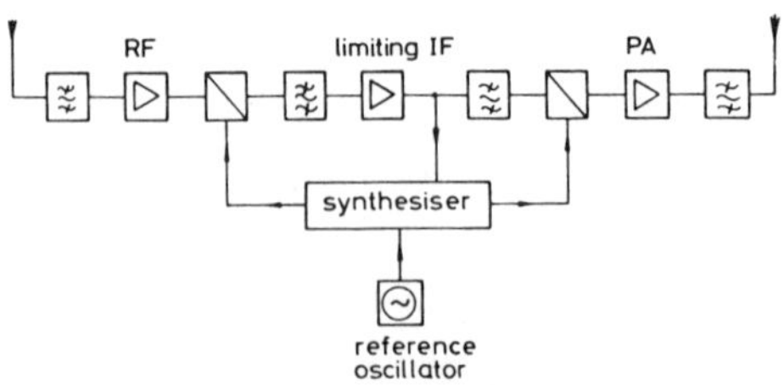

Fig. 1. Non-demodulating Repeater

7.7. Quasi-Synchronous Single Sideband Schemes.

Very little work has been done to evaluate single sideband (S.S.B.) modulation techniques for area coverage, presumably because as a modulation method S.S.B. has not yet found favour with mobile radio design engineers. The only scheme to be evaluated was on a purely experimental basis and used three conventionally spaced and controlled sites with the standard few Hertz offset as used with Q.S.A.M. schemes. The equipment employed produced standard S.S.B. transmissions and the mobile receiver used a manual clarifier. The overall performance with speech transmissions was very satisfactory even in low signal level overlap areas.

7.8. Return Path.

All the techniques so far described have been for the outgoing path to the mobile. However, all the main transmitter sites in an area coverage scheme will also be equipped with receivers, the audio output of which will be linked back either by landlines or radio links to the central control point. An assessment will then be made as to which receiver provides the best quality signal for use by the fixed operator, providing an improvement in the recieved signal from the mobile and portable stations.

7.9. Area coverage using satellites.

Clearly a satellite can provide coverage over all of the U.K. very easily. Therefore frequency reuse is naturally extremely limited. Although quite serious consideration is being given to satellite communications for mobile radio, it seems unlikely that it will replace any land-based systems in the immediate future, due to the cost of a satellite system and the excessively large "footprint" such a satellite possesses for a country as small as the U.K. The cost could be brought down by using existing sattelite capacity in a "piggyback" fashion, but the cost is likely to be prohibitive for mobile radio.

7.10. Conclusion.

Area coverage techniques were devised initially to provide cover on a county basis for police forces, and all the developments and refinements that have occurred since have been concerned with that and very similar requirements.

For voice and low speed analog data systems it would appear that there is an adequate range of techniques available for virtually all requirements. However, any widespread introduction of high speed data communication systems will introduce a necessity for much higher quality consistency of cover than is needed for speech communication. It is with this requirement in view that the more recent developments, specifically Sideband Diversity and Synchronous Area Coverage have been developed.

7.11 References.

Spaced Carrier and Other Mobile Radio Techniques.

(1) Brinkley J.R. "A Method of Increasing the Range of V.H.F. Communication Systems by Multi-carrier Amplitude Modulation." J.I.E.E. 1946, 93, Part.III, pp 159-166.

(2) Holbeche R.J. "Area Coverage Techniques for Mobile Radio." Communications 74 Conf., Brighton, June 1974, Paper 4.1.

Quasi-Synchronous Area Coverage.

(3) Carter D. "Quasi-Synchronous Operation of A.M. Transmitters." I.E.E. Int. Conf. Communications Equipment and Systems, 1976, pp 90-93.

(4) Gosling W., Holbeche R.J. "A Feasibility Study for a Voice Plus Data Radio System of the Future." I.E.E. Int. Conf. Communications Equipment and Systems, 1976, pp 121-127.

Quasi-Synchronous F.M. or Simulcast Techniques.

(5) Gray G.D. "The Simulcasting Technique: An Approach to Total-Area Radio Coverage" I.E.E.E. Trans. on Vehicular Technology, Vol. VT-28, No. 2, May 1979, pp 117-125.

Sideband Diversity.

(6) Allen G., Holbeche R.J., Gosling W. "An Evaluation of a Sideband Diversity Technique for Data Transmission on the Forward Path in a Mobile Radio Area Coverage Scheme." Radio and Electronic Engineer, Vol. 49, No. 10, October 1979, pp 521-529.

(7) Gosling W.,Martin J.D., Holbeche R.J., Allen G. "Sideband Diversity : A New Application of Diversity Particularly Suited to Land Mobile Radio." Radio and Electronic Engineer, Vol. 48, No. 3, March 1978, pp 133-139.

Synchronous Area Coverage.

(8) Holbeche R.J., Morcom C.J. "Synchronous Area Coverage for U.H.F. F.M. Portable Radio Telephones." I.E.E. Proc., Vol. 129, Pt. F, No. 4, August 1982, pp 249-253.

Quasi-Synchronous S.S.B. Operation.

(9) Gosling W., Holbeche R.J. "Home Office Speech Plus Data Project - Final Report." University of Bath, School of Electrical Engineering, July 1977.

Chapter 8

Data systems

P.J. Mabey

8.1 INTRODUCTION

Digital transmission is an essential part of modern mobile communications systems. It is used for high capacity radiopaging systems, for controlling trunked radio channels, and for public mobile telephony. The introduction of digital transmission is a result of an increased demand for mobile radio and the need to use the radio spectrum more efficiently.

Digital transmission is also required for communicating data. Data communication is superior to speech communication in some applications, because data messages can be sent more quickly and reliably and with less operator involvement. New facilities are possible, such as vehicle printers and direct access to a central computer, which can enable a vehicle fleet to operate more efficiently.

Unfortunately mobile radio has to operate under extremely unfavourable conditions, suffering from the combined impairments of ignition noise interference from nearby vehicles, signal fading caused by multipath propagation and shadowing, and co-channel interference which results from sharing the radio channel locally and re-using the channel in another part of the country.

In this chapter some of the key features of mobile digital transmission are reviewed.

8.2 DATA MODULATION

Initially, signalling over mobile radio was performed using audio frequency tones, for simple applications such as tone coded squelch and for selective calling. Tones can be decoded reliably at very low signal to noise ratios even in the presence of fading, because the tone duration is longer than a signal fade. For example a selective call lasts typically 200 ms using a sequence of five tones each lasting 40 ms.

Tone signalling is too slow and inefficient for many of the more recent applications requiring signalling. The data rate of tone signalling systems is typically about 90 bit/s, which is very inefficient in an audio bandwidth of 3 kHz and an RF bandwidth of 15 kHz at UHF, and is similar to the information rate of speech.

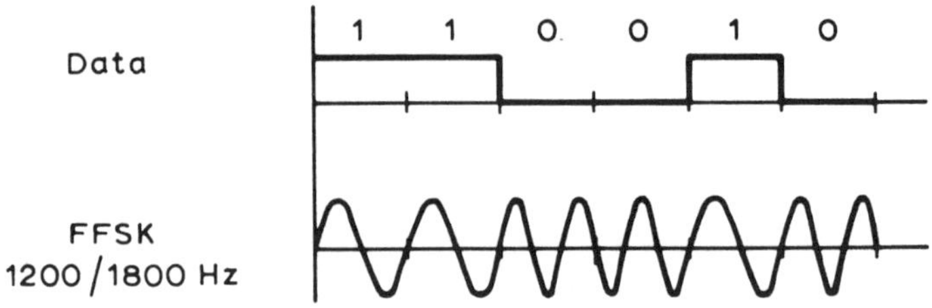

Fig.8.1. Fast Frequency Shift Key Modulation

Higher data rates have been obtained by using various binary modulation schemes in which an audio tone, termed a 'subcarrier', is frequency or phase shift keyed. A modulation method commonly used for digital transmission is Fast Frequency Shift Key (FFSK) at 1200 bit/s, in which a binary '1' is represented by one cycle of 1200 Hz, and a binary '0' is represented by one and a half cycles of 1800 Hz (Fig.8.1).

Subcarrier and tone modulation schemes are attractive because they occupy the audio frequencies and so are compatible with existing radios designed for speech

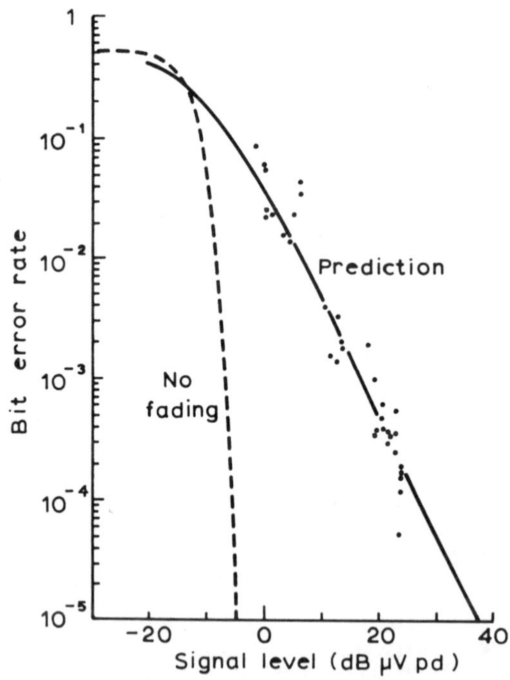

Fig.8.2 Field measured bit error rates for a moving vehicle, 462 MHz, 1200 bit/s, PSK subcarrier

communication and with telephone lines. If the speech circuits of a radio are bypassed then higher data rates are possible by modulating the RF carrier directly with the data signal. There are particularly efficient forms of direct frequency modulation, such as Tamed Frequency Modulation (Muilwijk (1)) which enables a data rate of 16 kbits/s with 25 kHz channel spacing.

8.3 BIT ERROR RATES

Errors occur when the received signal level drops to the front end noise level of the receiver. The bit error rates measured in a moving vehicle are high, typically in the range 0.001 to 0.0001 (Fig.8.2).

It is possible to predict the bit error rate from a bench measurement of bit error rate with a steady signal level (French (2)) (Fig.8.2) by averaging the bit error rate as the signal level varies with the Rayleigh and lognormal characteristics of mobile radio propagation.

The number of errors caused by ignition noise depends on the received signal level, the bit rate and the carrier frequency. At 1200 bit/s 165 MHz, about the same number of errors are caused by fading into the receiver front end noise as by ignition noise in heavy traffic (2), although at UHF ignition noise is not a significant source of errors at 1200 bit/s. Ignition noise becomes more detrimental as the data rate is increased or the carrier frequency is reduced.

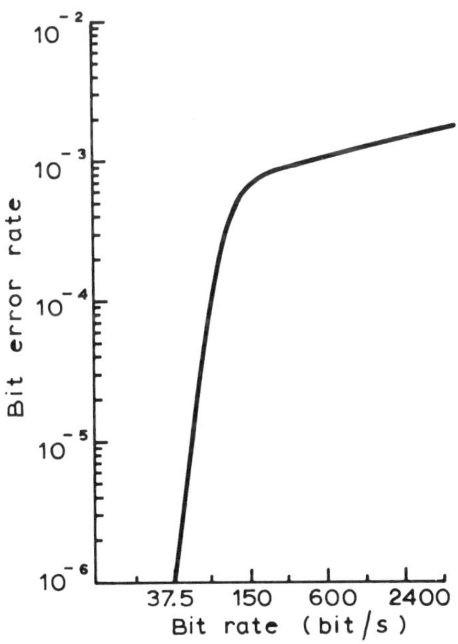

Fig.8.3 Error rate dependence on bit rate, moving vehicle

8.4 BIT RATES

The bit error rate experienced in a moving vehicle will depend, among other things, on the signalling speed (Fig.8.3) (French (3)). At a very low signalling speed the duration of a bit is much longer than the duration of the fades so there are very few errors. The error rate rises steeply as the speed is increased until the fades span many bits, in which case the bit error rate depends mainly on the proportion of time for which the signal is faded. Operation at a high bit rate, with sufficient redundant coding to detect or correct the errors, enables the highest data transfer rates.

In contrast, a low bit rate will give a greater sensitivity (Fig.8.4) by using a lower bandwidth, resulting in a slightly greater range for stationary vehicles receiving a steady signal level. For example, increasing the sensitivity by 2 dB increases the range by 12 %, using the commonly accepted inverse fourth power law for the path loss.

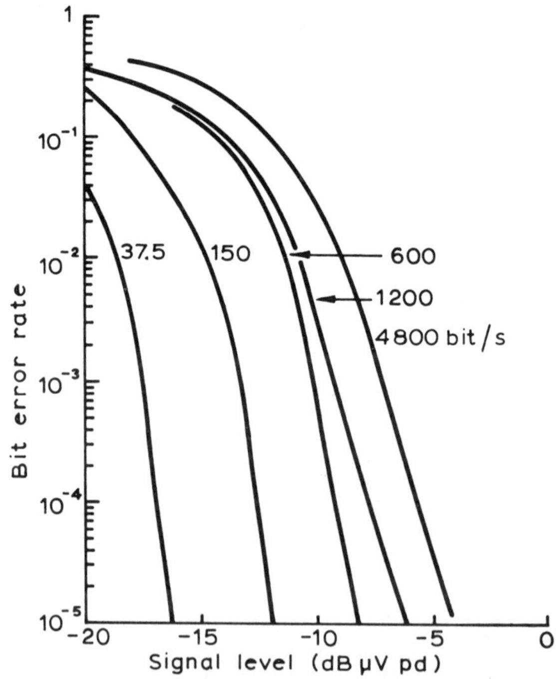

Fig.8.4 Error rates without fading, various bit rates

8.5 QUASI-SYNC

Quasi-synchronous transmission (French (4)) from a multiple transmitter is used to improve coverage where a single transmitter is inadequate. There are several constraints which must be observed to ensure successful data communication in the overlap regions, where a mobile receives roughly equal signal levels from two or more transmitters.

Transmitter sites are commonly fed via land lines, and the delays over the lines must be equalised to within acceptable limits. For example, Fig.8.5 shows that with direct frequency modulation at 1200 bit/s the differential delay must be less than about 0.3 of a bit period, 250 microseconds. The delay equalisation is more stringent for subcarrier modulation schemes and for the higher bit rates. It is important that the modulating signal transmitted from each site is the same. Any subcarrier frequency offset introduced by frequency division multiplexing on the landlines must be removed before transmission.

A stationary mobile can experience a problem in the overlap region at locations where it receives signals of a similar level. The resultant signal beats at the difference frequency of the transmitters, and can beat below the noise level of the receiver, causing errors (Fig.8.6). The difference frequency must suit the signalling speed and the coding used, so that there is a good chance of receiving a message between beats, or so that the code can correct an error burst occurring during a beat. For example, with a 64-bit code transmitted at 1200 bit/s, a carrier offset of 10 Hz would produce a beat every 20 code words.

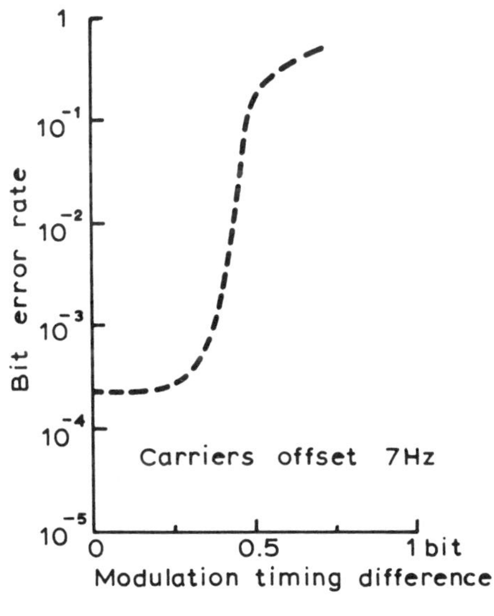

Fig.8.5 Error rate dependence on differential delay

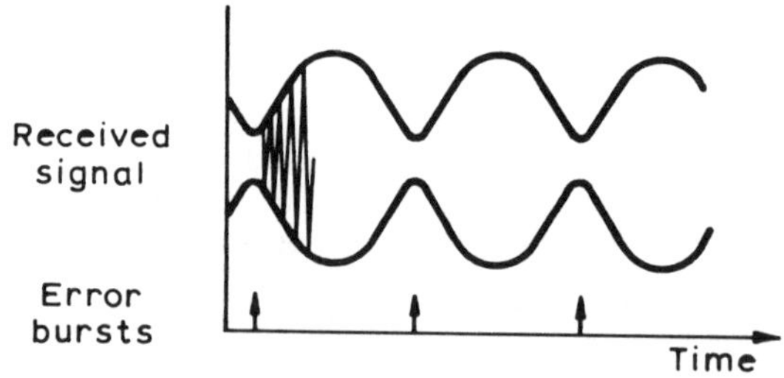

Fig.8.6 Signal envelope for stationary receiver

8.6 CODING

With the high error rates that occur with high data rates, error control coding is essential to ensure that messages are received successfully and that false messages are rejected.

8.6.1 Codes

We shall consider block codes, in which data is encoded by dividing the information bits into groups of k bits and adding r bits of redundancy to each group, derived by parity checks on the information, to form an n-bit block or code word. It is convenient to refer to an (n,k) code, for example (31,21). A code may be used for detecting errors, correcting errors occurring randomly within the code word, or for correcting clustered errors occurring in bursts. There are three possible outcomes from decoding a received code word. The message may be delivered correctly to the data sink, with probability Pc, called the success rate; errors may be detected but not corrected, with probability Pd, so the message cannot be delivered; the decoder may decode the message erroneously and deliver a false message with probability Pe. A code and decoder must be chosen carefully to ensure a good compromise between success rate, protection from false messages, and added redundancy.

The number of bit positions in which two code words differ is called the distance between code words, and the minimum distance for a code is denoted d. If we use an error detecting decoder then when fewer than d errors occur we can be sure of detecting them. But if d or more errors occur there is a risk of a false message. If we use an error correcting decoder then we can be sure of successfully correcting $t=(d-1)/2$ or fewer errors, and if more occur there is a risk of a false message.

8.6.2 Code Performance

To assess the performance of a code we need to know how often a code word is corrupted by errors and how often the error detecting or correcting ability of the code is exceeded. This can be found by analysing field measured bit error patterns (Mabey (5)). The error pattern is divided into blocks of bits and the number of errors in each block is counted. The results of this are plotted as the probability P(≥m,n) of finding m or more errors in a block of n bits. Fig.8.7 illustrates a measured P(≥m,n) distribution for a bit error rate of 0.01 and shows, for example, that 20 % of 63-bit blocks contained errors. A (63,39) code can correct up to four bit errors, and 4 % of blocks contained five or more errors.

Code success and false rates can be calculated from P(≥m,n) distributions using the following equations (Lucky et al (6)).

$$Pc = 1 - P(\geqslant e+1, n)$$

$$Pe = P(\geqslant d-e, n) \cdot 2^{-r} \cdot \sum_{i=0}^{e} C_i^n$$

where e is the number of errors the decoder is designed to

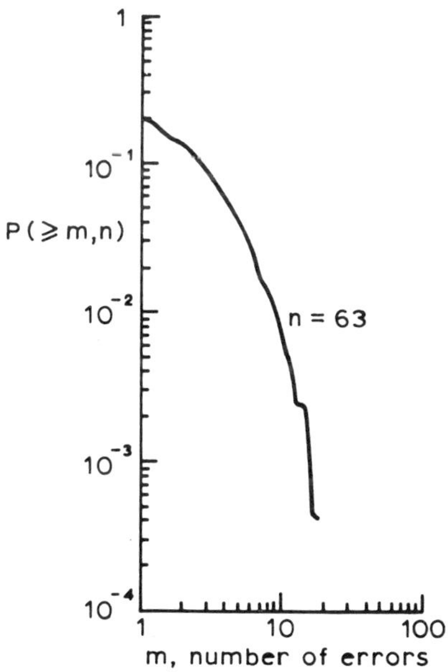

Fig.8.7 Typical P(≥m,n) distribution

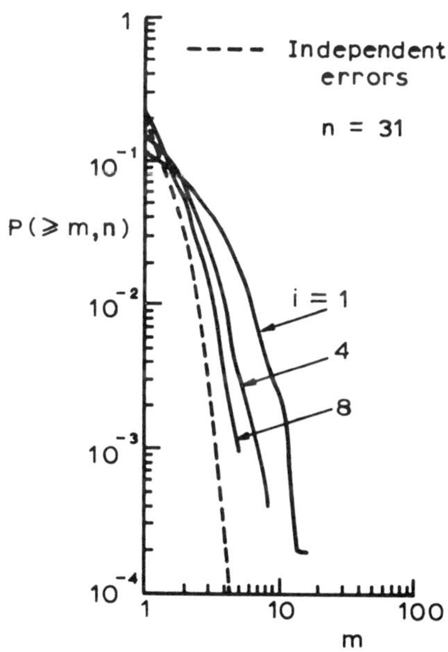

Fig.8.8 P(≥m,n) distributions for interleaved words

correct, etc. Similarly, analysis of error burst length distributions can be used to evaluate burst correcting codes.

8.6.3 Interleaving

Bit interleaving can be used to improve performance when more than a single code word is transmitted. It is a method of dispersing the errors that occur in clusters when the received signal level fades, and which are likely to exceed the correcting capability of a code. Before a message is transmitted the bits from several code words are interleaved. Then when an error burst occurs, the errors will be shared among the interleaved code words and a less powerful code is required to correct them. The process may be visualised as assembling i code words as rows of an nxi array and transmitting the bits by scanning the columns. The number of code words i which are interleaved is called the interleaving degree. The interleaving process does not involve adding more redundancy.

Fig.8.8 illustrates how the likelihood of multiple errors in a 31-bit code word is reduced by interleaving to various degrees. The improvement gained depends on the code word length, the amount of error correction performed and the fade rate. However, care must be taken not to degrade performance. For example, at slow speed a long error burst normally corrupting only a single code word would be distributed by interleaving and might exceed the error correcting ability in several code words.

8.7 UK STANDARD

In the UK the Electronic Engineering Association (EEA) and the Department of Trade and Industry radio regulatory department have specified a modulation scheme and a binary data format for mobile data transmission (7,8). The standard is described in Code of Practice MPT1317. The modulation scheme is 1200 bit/s FFSK. The format is illustrated in Fig.8.9.

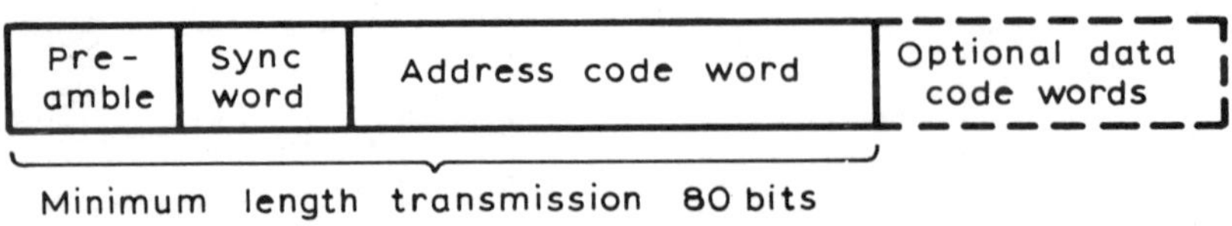

Fig.8.9 MPT1317 data format

Transmissions begin with a preamble of bit reversals 1010..10 so that the receiver data demodulator can acquire bit synchronisation. Every message begins with a 16-bit synchronisation word to enable the decoder to establish code word framing. Messages are transmitted in code words from a (64,48) code. The first code word of a message (called an address code word) contains addressing information and some data, and is sufficient for short message applications such as status reporting or trunked system call set up. For longer messages with more data, such as text, additional data code words are appended as required to accommodate the message.

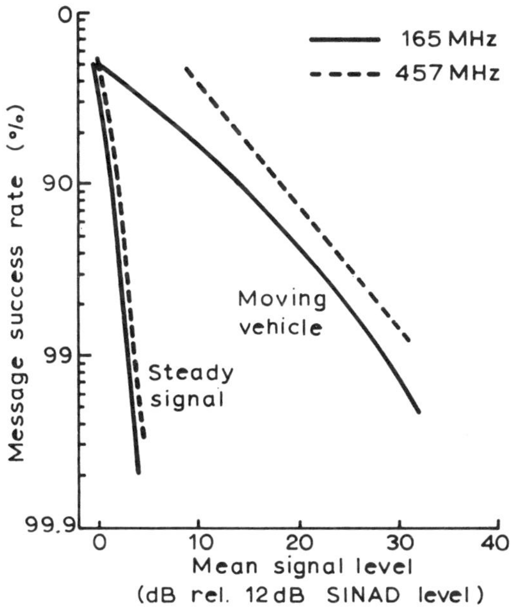

Fig.8.10 Message success rate for data format

Fig.8.10 shows the message success rate when short messages without data code words were transmitted over conventional FM mobile radio equipment with FFSK data modulation and a simple error detecting decoder. The steady signal curves were measured in the laboratory and apply to a stationary vehicle. The moving vehicle curves were measured during field tests under typical fading conditions. These results are used in the computer simulations of mobile data systems described in section 8.9.

8.8 DATA PROTECTION

Error correction can be used with the MPT 1317 data format. The amount of error correction that is acceptable is limited by the false call probability which degrades as more correction is incorporated. False calls can arise in several ways:

- False address word, in which errors corrupt an address code word causing it to be decoded as a different address word,
- Spurious call, in which random bits demodulated during noise, speech or interference mimic a

message,

- Misframed call, in which a synchronisation word and codeword are mimicked in the transmitted data sequences.

It is not possible to totally prevent false calls, but they can be made acceptably rare. Also, the acceptable false rate may depend on the direction of transmission (to or from a mobile) and on the particular message meaning, as some false messages will be more disruptive than others. Apart from the falsing protection inherent in the code, additional protection is derived from:

- Addressing. A mobile will accept only messages addressed to it.
- Context. Messages received out of correct protocol context can be rejected.
- Timing. Message transmission may be confined to designated time slots, and there would be a prescribed time limit for accepting responses to messages.

In addition it is necessary to consider the ways in which false data words can arise:

- False data word. A data code word may be corrupted into a different code word.
- Bit Slip. A burst of noise or interference may cause bit slip in the data demodulator so that data words are decoded misframed.
- Interference. Another transmission can overlap in time and substitute part of a message, so that the decoded message is a composite of wanted and interfering code words.

The falsing mechanisms outlined above must all be considered when designing a mobile data system.

8.9 RETRANSMISSION PROTOCOLS

An effective way of improving the reliability of data communication is to retransmit messages, or parts of messages, not received successfully. The choice of retransmission protocol will influence the maximum geographical range of operation, the number of vehicles which can be accommodated in a system, and the response time or delay experienced by users. For illustration we shall consider a simple protocol for communicating a message using the MPT1317 data format, and occupying ten data code words. The sender will retransmit the whole message, subject to a maximum number of attempts, until the recipient acknowledges that it has assembled the whole message successfully.

It is not sufficient to evaluate a protocol at a specified error rate. There are many features of mobile radio which should be taken into account and computer simulation is an important tool for doing this. For example, vehicles are distributed throughout a service area and so have a range of error rates. Some vehicles are stationary and receive a steady signal level, whereas others are moving and receive a fluctuating signal level. Mobiles may transmit and receive on different frequencies, and both frequencies must have a good signal level for successful communication. Some results of simulations (Mabey (9)) are presented below.

8.9.1 Comparison With Speech

A common requirement is that the limit of range for data communication is at least as good as the range for speech communication. To enable comparison we will define a reference level for speech communication.

Initially consider stationary vehicles receiving a steady signal level. For private mobile radio users with two frequency operation we require ≥12 dB SINAD on both the transmit and receive frequencies for a conversation to be possible. (The 12 dB SINAD threshold is a minimum; speech quality is poor and there is no margin for environmental noise. Often, a higher threshold would be used in practice). We shall assume that the mean received signal level is equal for both frequencies. The probability of ≥12 dB SINAD on both frequencies is plotted in Fig.8.11 curve c), for uncorrelated Rayleigh fading and correlated lognormal shadowing (with a standard deviation of 6 dB, generally accepted as typical) on the two frequencies. For a

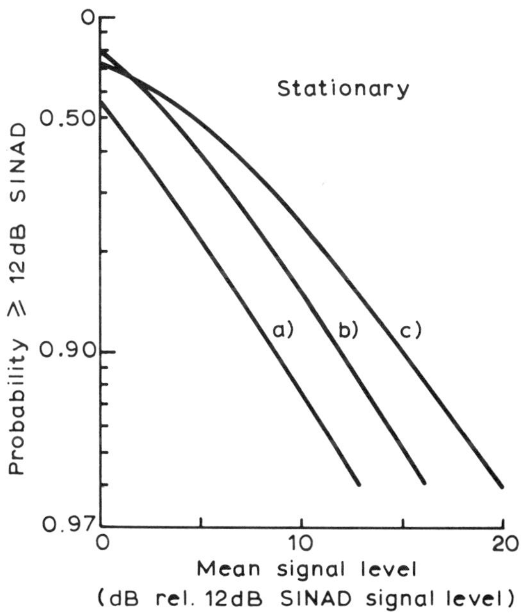

a) single frequency, fading only
b) two frequencies, fading only
c) two frequencies, fading and shadowing

Fig.8.11 Signal level for speech communication

conversation to be possible at 90 % of locations at the edge of the service area the mean signal level at the edge must be 15 dB above the 12 dB SINAD level. Therefore we define the normalised range to be unity at 15 dB above the 12 dB SINAD level. If we had ignored the shadowing variations or considered only one frequency, then Fig.8.11 curves a) and b) show that we would have under estimated the required signal level.

The mean signal level required for speech communication in a moving vehicle is more difficult to quantify. However, at a mean signal level of 15 dB above 12 dB SINAD, the signal level falls below the 12 dB SINAD threshold for only 6% of time. We will assume that 15 dB is also adequate for moving vehicles.

For comparison with speech, we shall define the maximum range for data communication as where the data transaction success rate has fallen to 90 %.

8.9.2 Range

The maximum range for data communication obtained from the computer simulations of the ten data code word message is shown in Fig.8.12. and is normalised to be unity at the limit of range for speech. The range for stationary vehicles is

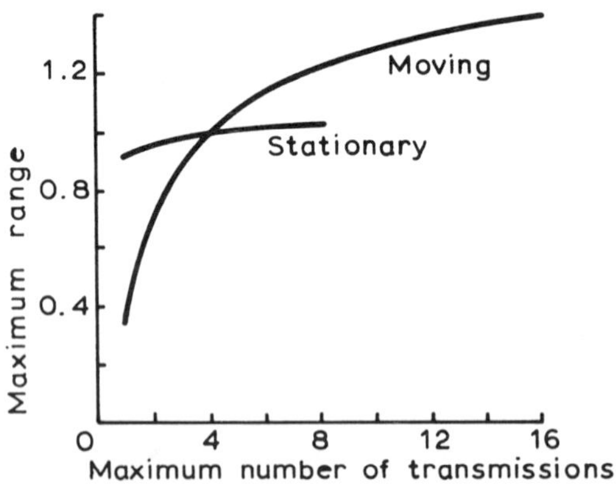

Fig.8.12 Maximum range

comparable to the range for speech communication and cannot be significantly improved by allowing many retransmissions. Stationary vehicles normally have either a very good signal level or an unuseable signal, such that few vehicles benefit from repeats. Moving vehicles can have a long range because they can use the retransmissions to decode the code words corrupted during fades in the initial transmission.

8.9.3 Throughput

Throughput is defined as the proportion of time used for transmitting successful new information bits, and is

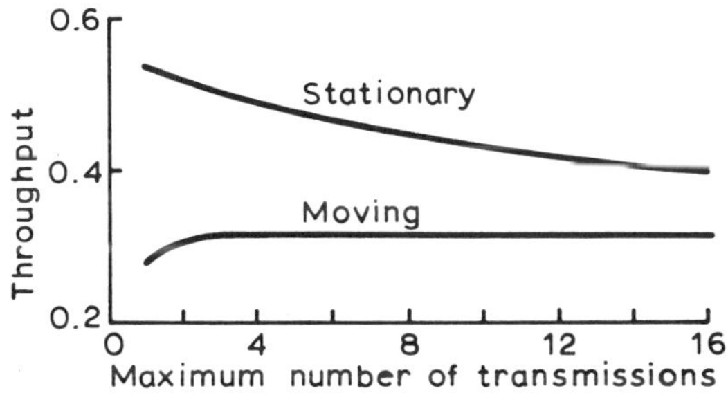

Fig.8.13 Throughput

evaluated as the average for the whole fleet distributed throughout the service area. Stationary vehicles achieve a higher throughput than moving vehicles (Fig.8.13) because they require fewer retransmissions, but the throughput decreases as more transmissions are permitted. Vehicles stationary in a signal fade will require the maximum number of retransmissions permitted, and will still not succeed, whereas nearly all moving vehicles are successful with three or fewer transmissions.

8.9.4 Delays

Moving vehicles experience longer delays in communicating data messages than stationary vehicles and have a broad distribution of delays. Fig.8.14 shows the histogram of delays at the edge of the service area for a longer message with 100 data code words. For the simple retransmission protocol described, and with a retransmission limit of eight times, the mean delay is 24 seconds, but 20 %

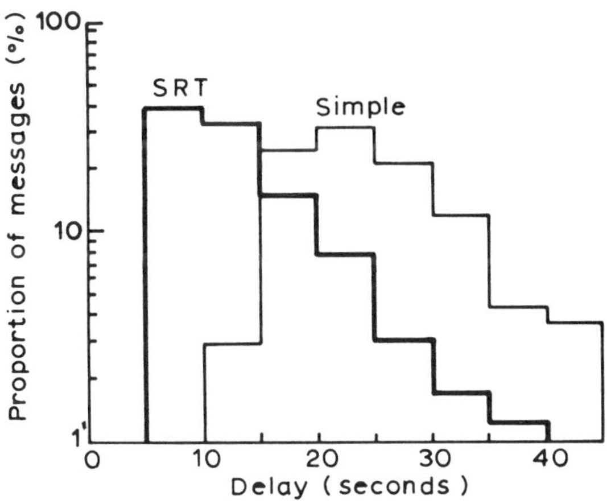

Fig.8.14 Delay histograms

of vehicles suffer delays exceeding 30 seconds. A selective retransmission protocol (SRT) which avoids repeating the code words already decoded successfully reduces the mean delay to 14 seconds, and only 4 % of vehicles suffer delays exceeding 30 seconds.

8.10 CONCLUSIONS

The mobile radio channel is very poor and presents many problems to digital transmission. Reliable data communication requires the careful application of coding and protocols. Computer simulation is an important tool for predicting the performance of systems which are too complex for calculation and impractical to measure. Digital transmission will make an important contribution to spectrum efficiency, and its use is increasing as the demand for mobile radio grows and as users come to expect more sophisticated facilities.

REFERENCES

1. Muilwijk, D., 1979, 'Tamed Frequency Modulation - a bandwidth saving digital modulation method, suited for mobile radio', Philips Telecommunications Review, 37, 35-49.

2. French, R.C., 1978, 'Error rate predictions and measurements in the mobile radio data channel', IEEE Trans. Veh. Tech., VT-27, 110-116.

3. French, R.C., 1980, 'The mobile radio data channel', Proc. 1980 Int. Zurich Seminar on Digital Communications, D1.1-D1.9.

4. French, R.C., 1980, 'Common channel multi-transmitter data systems', Radio and Electronic Engineer, 50, 439-446.

5. Mabey, P.J., 1978, 'Mobile radio data transmission - coding for error control', IEEE Trans. Veh. Tech., VT-27, 99-109.

6. Lucky, R.W., Saltz, J., and Weldon, E.J. Jr., 1968, 'Principles of Data Communication', New York: McGraw-Hill.

7. 'EEA preferred data transmission system for land mobile radio using a binary format', 1982, Electronic Engineering Association, London.

8. 'Transmission of digital information over land mobile radio systems', 1981, Code of Practice MPT1317, Home Office Radio Regulatory Department, HMSO, UK.

9. Mabey, P.J., 1982, ' Predicting the range and throughput of mobile data systems', IEEE Veh. Tech. Conf., 370-374.

Chapter 9

Paging systems

A.K. Sharpe

9.1 INTRODUCTION

Mobile radio by definition is for communication with people on the move. In the fullest sense this is not limited to vehicular movement but personal communication. Two people are required for any form of human communication but for good communication everyone needs to be contactable. Because of the finite frequency resource, propagational characteristics of radio and technology limitations many different land mobile radio systems have been proposed.

Paging is one such system which attempts to solve the "all times, all places, all people" communications problem. We will see how, with very economic use of the r.f. spectrum, paging is able to keep very large numbers of people in touch with each other at essentially low cost with small, light-weight, unobtrusive, bodyworn equipment. To achieve this paging is fundamentally a one-way calling system and relies on the availability of a wire-line telephone network to support the communication process. From simple beginnings where in building public address systems were replaced and improved by personal paging systems, very small sophisticated Pagers operating in all frequency bands with powerful message capabilities, and supported by nationwide transmitting and control infra structures, have been developed.

9.2 ON-SITE PAGING

In the mid 1950's the problem of calling doctors and nursing staff silently without distressing patients was addressed by the radio technology of the day. The first paging system to be introduced, circa 1957, by the Multitone company, used low frequencies in the range 30KHz to 50KHz and was conveniently restricted in range by the use of inductive loop techniques. Each receiver was simply tuned to its own carrier frequency. As more users were required frequency modulation of a 40KHz carrier was introduced.

Inductive loop systems are still widely used today and are favoured by regulatory authorities because of their excellent spectrum conservation properties. However the cost of installation, extension and coverage prediction makes such system unattractive in comparison to radio systems. Typical Inductive Loop systems operate with approximately 1.5A loop current from 25W drivers at 40KHz and have an inductance of approximately 1.5mH over 200m to 500m with standard 0.5mm flex. Pagers have typically a 10μA/M to 30μA/M sensitivity with a frequency deviation of ±4KHz.

On-site Paging rapidly expanded during the 1960's and 1970's as frequencies in the 27MHz to 42MHz and 470MHz bands were allocated. Channel separations moved quickly from 25KHz to 10KHz in the v.h.f. bands, with amplitude modulation, and remained at 25KHz in the u.h.f. band, with frequency modulation. Until about 1977 the signalling systems were based on sequential tones in the frequency band 500Hz to 3KHz, however today there are many different proprietary digital signalling systems on the market with no compatability existing between different manufacturers equipment. Transmitters are typically in the 1W to 5W range, and pager sensitivities are in the range 10μV/m to 100μV/m measured in the best direction, off the body.

On-site Paging is essentially a low capacity system of communication with typically 20 pagers per system and a maximum of 3000 pagers to a system. Speech is often transmitted and digital displays are used. The provision of answer-back facilities with integral transmitters radiating in the 146MHz to 169MHz band, 20mW ERP, from antennas with effective gain of -6dB, is often a part of an on-site paging system. The use of radiating cables and the integration of intercom, telephone and alarm systems all form part of an installation.

A U.K. specification for such products has been issued. See MPT1305

9.3 WIDE AREA PAGING

On-site Paging, with a typical range of less than 2Km, grew as an extension of the internal telephone, public address and intercom system business. However applications for the facilities offered in the on-site situation were soon found for much wider area systems requiring greater range. Fireman's call-out and medical answering services are examples. The early systems of this type required limited address capacity but excellent sensitivity to achieve maximum range from the relatively simple and, by necessity because of the low capacity, low cost fixed transmission networks. The early wide area paging systems, unlike on-site systems, were seen as an extension of the mobile radio business, and this was reflected in the companies entering the market, the specifications and the frequency ranges.

The concept of offering a paging service more widely than within a private or professional organisation was first introduced in North America in 1963, and then quickly in other countries, notably in Europe and Japan. Figure 1 shows the significant milestones in the growth of wide area paging. Some of these systems are of considerable importance and will be considered in detail, however, what is not revealed is the multiplicity of incompatible systems that have been and currently are in operation.

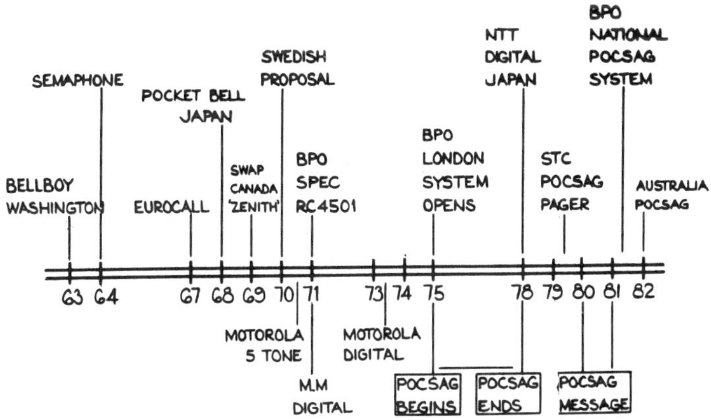

Fig. 1 Paging Milestones

9.4 SIGNALLING METHODS

The early systems used sequential tone signalling and the latter digital signalling. Of the tone signalling systems, a division can be made between those with typically two sequentially transmitted audio frequencies and those with typically five or more. All the tone frequencies used to modulate the r.f. carrier are in the 100Hz to 3000Hz audio band. In order to achieve an acceptable address capacity with only two sequential tones, a high level of f.d.m. is used with a choice made from 40, or in some cases 70, frequencies in that band. Thus, very high Q filters are required in order to identify the narrowly separated frequencies, high stability circuits are required in both transmitter and pager and the r.f. carrier must be modulated by each tone for a duration related to the Q of the decoding filters. Typically two tone paging systems have an address capacity of between 1000 and 3500, and a calling rate of between 1 and 0.3 calls per second dependent upon the range of frequencies chosen and the possible use of battery economiser techniques. Sensitivity improvements over other methods of signalling are in the order of 6dB for two tone systems because of the small noise bandwidth resulting from the use of high Q filters.

In order to relax some of the critical stability requirements necessary for the two tone systems and to give increased address capacity coupled with increased calling rate, what is sometimes called 'decimal digital' signalling was introduced. Five or more sequentially transmitted tones are chosen from an effective library of 10 frequencies giving a base 10 system for addressing. In fact, more than 10 frequencies are identified to allow for the tone repeat situation, group call and other additional special functions. While there is, in general, only one preferred series of tone frequencies used for two tone systems, there are many used for five, or greater, tone systems. The EIA/Motorola five tone system is based on an arithmetic progression of frequencies with a separation of 15Hz allowing constant Q techniques to be used in filter design, and relatively short tone durations, each of 33mS. This gives a calling rate of 5 calls per second and an address capacity of 100,000 is achiev able. Other five tone paging signalling systems are designed to share the same signalling as mobile radios and provide what is often known as 'overlay' paging. This means that ZVEI, CCIR and EEA signalling systems are employed. A very significant sequential tone system in use in Europe today is Eurosignal. Tone frequencies are within the range 313.3Hz to 1153.1Hz based on a geometrical progression. Each tone is transmitted for 100mS, and calls, each of 6 tones, are separated by 200mS of the highest frequency tone (1153.1Hz) called a 'line clear tone'. Thus the calling rate is 1.25 calls per second and the address capacity with base ten systems is 1×10^6 addresses but with a possible base of 14, increasing the address capacity to 7.5×10^6. [11] [12]. Figure 2 illustrates the Eurosignal structure.

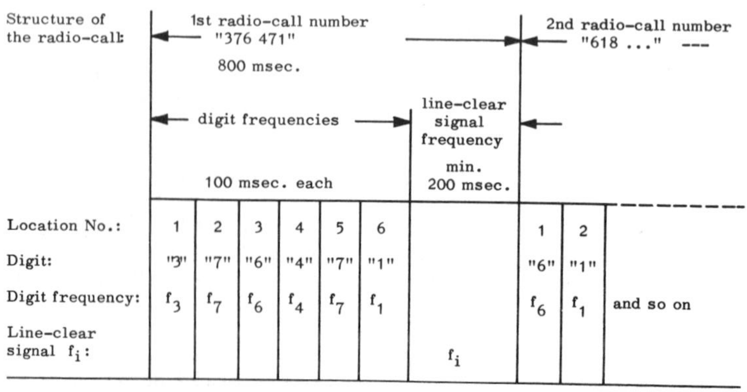

Fig. 2 : Euro-Signal Format

The first truly binary digital paging system was introduced in 1970, in Canada, and subsequently aroused great interest in such methods of signalling. The trend towards digital signalling techniques has not always been fully understood because digital signalling does have some disadvantages. Notably it will in general offer inferior sensitivity, and thus range, because of a greater noise bandwidth than tone signalling. This is especially true if equal or greater address capacities and calling rates are to be achieved. However, as the cost of digital technology falls, and the dual requirement for large address capacity and high call rate rise to meet the requirements of, at first, very large city wide systems and now nationwide systems, digital signalling has come into its own. Digital paging systems also have the advantage that additional functions and features can be easily provided eg. multiple pager addresses and text messages. Such features are among the reasons on-site paging system manufacturers have introduced digital signalling. Before examining some of the more common digital signalling codes, it is important to identify the two different methods of r.f. carrier modulation in use. Direct modulation of an r.f. carrier with an NRZ bit stream has the advantage of a high modulating deviation to modulating frequency (data rate) radio. The disadvantage is that d.c. disparity results and thus the modulation spectrum is centred on the low frequencies where, in typical PMR and line systems, significant group delay, attenuation and distortion is encountered. The alternative modulation technique used in on-site systems, PMR systems and some wide area paging systems is to shift the modulating spectrum away from d.c. so that the low frequency signal corruption is overcome. The provision of a sub-carrier or techniques such as Manchester Bi-valent coding, on which a binary one is represented by a mark-and-space, and a binary zero is represented by a space-and-mark, achieve this with some success. Such bi-phase FSK systems have the disadvantage of requiring twice the base bandwidth of direct NRZ FSK modulation for the same bit rate. However, all digital systems are wide band and thus subject to distortion. Thus great care in system design is necessary; transmitters specially designed for paging are often required. A summary of the essential differences between the three common signalling methods is given in Table 1. It is worth noting that a UK company , Consumer Microcircuits Limited, has developed a decoder that allows messages to be transmitted using sequential tone techniques.

	SENSITIVITY	CALLING RATE	CAPACITY	MESSAGE	COST
TWO TONE	VERY GOOD	POOR	POOR	VERY POOR	MODERATE
FIVE TONE	GOOD	MODERATE	GOOD	MODERATE	MODERATE
DIGITAL	MODERATE	VERY GOOD	VERY GOOD	VERY GOOD	VERY GOOD

TABLE 1. Paging Systems Comparison

9.5 POCSAG

From Figure 1, it can be seen that the first wide area paging systems appeared in North America and Japan in the early 70's and during the mid seventies digital paging systems were developed by some of the major companies, the Swedish PTT and the NTT. However, during this period one of the most adventurous single channel national paging systems in the world was designed and planned in the United Kingdom by the British Post Office.

System Wide Area Paging (SWAP) was initially proposed in which numerous proprietary digital paging codes could be accommodated by computer controlled Paging Terminal Equipment. Such a system was commissioned for public use in London in 1976. However, it was soon realised that a transmission system combining both simultaneous and sequential techniques was necessary to optimise traffic handling for total country coverage and that because of uneven population distribution a single paging code was also necessary if channel utilisation and occupancy was to be maximised.

The British Post Office realised it was important to select a paging code and a signalling format that would meet performance characteristics they could from experience clearly identify and also be internationally accepted.

It was with this aim and purpose that in December 1975 the Post Office Code Standardisation Advisory Group (POCSAG) was established. Under the enthusiastic and powerful leadership of Mr. R. Tridgell, 36 engineers representing 13 major paging companies, and manufacturing assqciations met 16 times over a period of two years. Some 68 technical pagers were presented and results from research work around the world considered.

The properties of a code ideal for large national or major city use were initially agreed and identified in Table II.

TABLE II. Ideal Code

Code Capacity	:	10^6 Pagers each with 4 addresses
Paging Rate	:	6 Calls per second
Radio Frequency	:	Unrestricted
Radio Channels	:	25KHz/12.5KHz
Modulation	:	Choice to optimise system performance
Transmission Mode	:	Sequential and simultaneous
Adjacent Transmitter Frequency offset	:	Practical and maintainable at low cost
Line Transmission	:	Dedicated with standard modems
Absolute Line Delay	:	100uS
Lost Call Rate	:	2% in limiting typical urban environment or along coverage boundary
False Call Rate	:	1 in 10^{-8}
Battery Economy	:	To be incorporated in code format
Paging Delay Time	:	2 minutes in all zones
Messages	:	Capability for numeric and alpha-numeric containing many characters. A single decimal character was not considered important
Code Compatibility	:	Cross calling with other codes to be avoided
Radio Practice	:	Not violated

Finally a code structure and format originally proposed by Philips Research Laboratories (PRL) and in particular Mr. P. Mabey, was accepted with some minor modifications as a code that satisfied all the requirements identified at the outset, while being non-proprietary and giving to no manufacturer an advantage.

The characteristics of this POCSAG paging code are fully described in the POCSAG report published in 1978 [6] and are summarised in Table III.

The code was initially evaluated by computer analysis at PRL using bit error distribution recorded in London under multipath, shadow and co-channel fading conditions. Finally the British Post Office also completed a comprehensive computer analysis using recorded bit error patterns at VHF obtained from a research project in Birmingham University. The results gave success and falsing rates under many different environmental conditions and in each case these were compared with the same parameters obtained from a known proprietary code. Field trials with a number of POCSAG pagers have also been conducted in London.

A novel and significant feature of the code format which was an original invention of the Philips Research Laboratories is the method used to batch code words within a frame concept. This is outlined in Figure 3.

TABLE III. POCSAG Code

Address Capacity	: 8×10^6 (2×10^6 pagers)
Paging Rate	: 15 calls per second
Message Rate Numeric	: 5 calls per second (10 digit message)
Message Rate Alpha	: 1.07 calls per second (40 character message)
Data Rate	: 512 b/s (+ 10ppm)
Channel Separation	: 12.5/25KHz
Modulation	: (NRZ) +4.5KHz (+2.5KHz) = 0 -4.5KHz (-2.5KHz) = 1
Transmission	: Simultaneous or sequential
Lost Call Rate	: 2%
Error correcting potential	: 2 bits random 5 bit burst per codeword
False Call Rate	: 10^{-8}
Battery Economy	: 18% battery use during normal transmission
Messages	: Yes (Alpha-numeric etc.)
Message Capacity	: Unlimited

Sixteen pager addresses are batched into eight unique time slots (frames) every 1.0625 seconds and in this way code capacity, signalling rate and battery economy are all optimised. Because Philips agreed not to apply for patent rights but offered the code for free international use, a high performance world standard paging code has emerged.

The POCSAG code format is based upon 32 bit codewords comprising a (31,21) BCH codeword with an appended overall even parity bit. Sixteen such codewords are concatenated in the form of 8 frames together with a leading synchronisation codeword to form a Batch. A call is determined by receipt of the synchronisation codeword together with an Address codeword received in a specified Frame, which Frame being determined from the pager address. Since a pager is only required to scan for its address in a known specified frame, once synchronisation is obtained, battery economy may be maximised by the pager switching off for all but its own frame. Batches may be concatenated ad infinitum, but on commencement of transmission, a preamble, consisting of data reversals is transmitted, to ensure that all pagers utilising battery economy are primed to receive synchronisation, and thence the first address codewords. Message codewords without limit may be transmitted immediately following the address of the destination pager, each message codeword replacing an address codeword within the batched structure.

The POCSAG code is designed for constant bit-rate transmission of 512 bits per second. Figure 3 shows the ...SAG signal format.

Each pager is allocated one of the eight available ...s per batch, and one combination of address codeword bits 2-19. Thus 21 bits identify a pagers response code. This is called a Radio Identity Code (RIC) and is normally marked on the pager case. Associated with each RIC are four addresses determined by the two function bits (20/21) transmitted. These address functions have been designated four alerting cadences and in display pagers indicate type of message formatting. A pager can be programmed with more than one RIC. Each RIC can be allocated up to four telephone numbers.

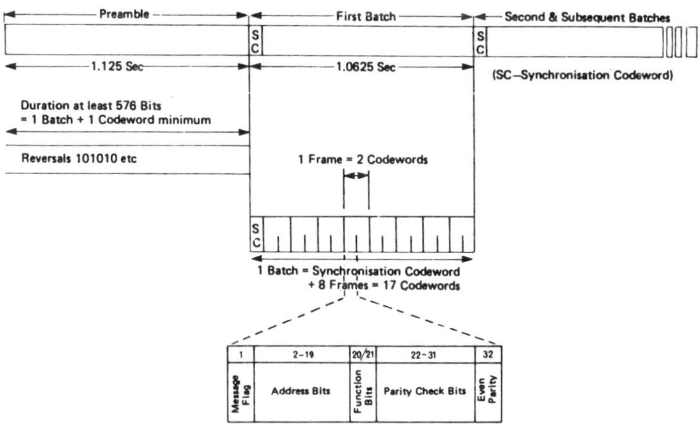

Fig. 3 : Diagram showing signal format (POCSAG)

A growing number of international manufacturing companies have announced POCSAG pagers and some of these have been evaluated by British Telecom. Many national PTT organisations have specified POCSAG for their nation-wide paging services. In February 1982 the POCSAG code was accepted by CCIR as a recommended No. 1 code for international paging.

The message capability of the POCSAG code identified in the summarised characteristics (Table III), is expected to have a marked impression on the development of paging. The structure of the code allows alpha-numeric messages to be transmitted together with the alert only calls on the same radio frequency. This means that simple numeric only message pagers, advanced alpha-numeric pagers and basic alert only pagers can be controlled from the same paging control equipment and introduced as they become available during the life of any national service. A POCSAG message codeword is like the address a (32,21) codeword. The first bit is a message flag, '0' identifies an address word and a '1' identifies a message word. The 20 information bits from 2 to 21 contain data and are formatted as 4 bit or 7 bit characters. Message codewords are always associated with the immediately preceeding address code and can be

Characteristics	MBS	GSC	POCSAG	NTT
Codeword Type	Kasami 26:16	Golay 32:12	BCH 32:21plus	BCH 31:16
Hamming Distance	3	7	6	7
Codeword Error Detection	Burst 10 errors Random 2 errors	Burst 11 errors Random 6 errors	Burst 11 errors Random 5 errors	Burst 15 errors Random 6 errors
Codeword Error Correction	Burst 5 errors Random 1 errors	Burst 5 errors Random 3 errors	Burst 5 errors Random 2 errors	Burst 7 errors Random 3 errors
Codeword Efficiency	$\frac{16}{26} = 0.62$	$\frac{12}{23} = 0.52$	$\frac{21}{32} = 0.66$	$\frac{16}{31} = 0.52$
Code Address Capacity	1×10^6	409,000	over 8×10^6	65,000
Number of bits per Address	52	60.5	32 (+3 implied) +2 overhead preamble	34.9
System bit rate (bit/s)	1187.5	300/600	512	200
Call rate per sec.	22.84	4.95	15.06	5.73
Battery Economy		No	Yes	Yes
Radio Channel in Use	FM Broadcast 88 to 104MHz	150 30MHz 450 930MHz	470MHz 150MHz	250MHz
In Service Since	1978 (Sweden)	1973 (USA)	1981 (UK)	1978 (Japan)

TABLE IV. Paging Code Comparison (Tone only)

concatenated to form any length of message. Messages are not restricted to frames or batches although sync words must be transmitted every 17 codewords in accordance with the POCSAG format of Figure 3. Messages are terminated by an address codeword which could be the address codeword begining a new message to the same or a different pager. Messages can be of any length. Address Function '00' identifies the begining of a numeric message formatted with five 4 bit hexadecimal characters per codeword. Sixteen characters are defined including 0-9, (), SPARE, U, :, Address function '11' identifies the begining of an Alpha message formatted with $2^{6}/{}^{7}$ seven bit ASCII characters per codeword. The full 128 ASCII characters are defined providing full upper and lower case capability and printer functions.

The wide interest in message paging was demonstrated by the increased international membership of POCSAG when the group was reconvened in 1980 and 1981 to draft a more detailed specification for message formation. The success of POCSAG must to a great measure be attributed to its Chairman Mr. R. Tridgell who has shown that international standards can be achieved if with skill the right ingredients are brought together.

Although there are other important digital paging codes, POCSAG is more fully described because it has the widest international acceptance. Three other important codes are compared in Table IV. Of these, the GSC code is supported by the largest manufacturer with a full range of products and is in volume service, notably in the USA. The NTT code is in high volume service in Japan and also sold by NEC in other countries. The Swedish MBS code is in use exclusively in Sweden, but has been seriously considered in other countries because it uses an already existing infra structure of broadcast transmitters. [9] Considerable rivalry has arisen between the GSC code, essentially American in origin, and POCSAG, a British code, resulting in claims and counter claims over their relative merits [7] [15].

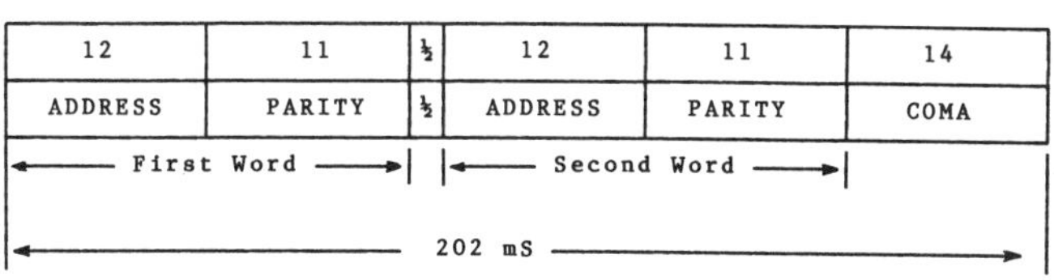

Fig. 4 : Golay Sequential Code Address Word

The GSC code has two bit rates 300 bits per second for address codewords and 600 bits per second for message codeword, with differing code structures for each type of codeword: Golay Sequential for addresses and interlaced-BCH for message codewords.

The format of the GSC shown in Figure 4 and compared in Table IV is for tone only address paging without battery economy. The interlacing technique of assembling the words in rows and transmitting in columns is designed to give improved burst error correction capability.

The relative differences in the vital parameters of digital signalling code structures, of lost call rate and false call rate, are only marginal and vary according to message length, speed of movement, orientation and environment of the pager. There is always going to be a slight trade-off between these parameters and the system parameters of calling rate, address capacity and battery economy techniques.

9.6 THE PAGING RECEIVER

The fundamental user requirement, irrespective of signalling method and systems infra structure, for a paging receiver is that it should be small, light, unobtrusive, low cost, robust, have a long battery life, and, of course, be 100% reliable under all conditions. To meet all these simultaneously has created some interesting challenges for pager designers. To offer reliable service the r.f. performance must be compatible with other mobile receivers operating in the same frequency bands. If a pager is to satisfy most of the specifications issued by national PTT organisations [1] [2], then its summarised performance must be:-

Frequency	138-174MHz
Channels	25KHz and 12.5KHz
Modulation	POCSAG - Radio Paging Code No.1 (CCIR 584)
Frequency Offset	±1010Hz
Frequency Stability	2ppm over 1 year
Sensitivity	10μV/m (8 position average on body on 30m test site)
Selectivity	-65dB at ±25KHz (or 12.5KHz)
Spurious Response	-70dB at ±50KHz
Intermodulation	-50dB unwanted signals + (25 and 50KHz)
Desensitisation	-70dB at ±25KHz
Spurious Radiation	2 nanowatts
Battery life	650 hours

The parameter that causes most difficulty is sensitivity, because it is adversely affected as size of pager and power consumption are reduced. As previously stated the signalling system also has a marked effect on sensitivity. For example typical two tone pagers would be expected to have a sensitivity of 4μV/m at 150MHz. The specification for Euro signal which is a 6 tone system with amplitude modulation on 87MHz calls for 2μV/m in any direction in free space, or on the body 8 position average of 2μV/m. Sensitivity is also subject to considerable dispute over acceptable and repeatable test methods.

Pagers have inefficient integral antennas and are normally worn on the body, as might be expected this has a marked effect on antenna performance. Some regulatory authorities and major operators thus specify thatmeasurements must be made with the paging receiver body worn, while others, because of the difficulties in obtaining repeatable measurements, specify free space sensitivity. These latter measurements have the disadvantage that an acceptable sensitivity figure for field coverage calculations is not obtained. The free space measurements can be made in a test cell or stripline while the bodyworn measurements need a 30 metre test site. Such test sites are costly to construct and to calibrate. The recently published IEC standard 489 Part 6 - Selective Calling - Receiver Decoder Equipment; RF Receiver Decoder Parameter Measurements, described a 30 metre test site together with a sensitivity test method which involves the use of a simulated man, an acrylic column of salt solution and requires an up-down technique of making measurements.

Most pagers today use a frame or loop antenna for frequencies above 100MHz, although some manufacturers prefer ferrite antennas. The effect of the body on antenna performance is shown in Figure 5. The body removes the two relatively large nulls but does attenuate the r.f. signal when the maximum body thickness is presented to the transmitting antenna. This effect is especially marked

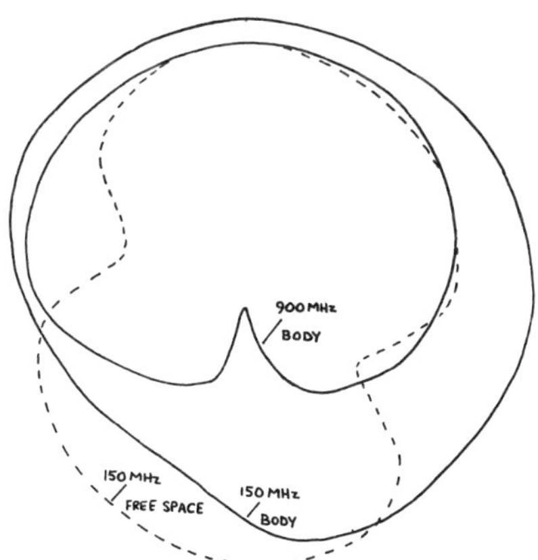

Fig. 5

at higher frequencies. Since the weakest field strength conditions are most likely to be found in buildings, it is argued that the field will always have a Rayleigh distribution, due to multipath effects, and that free space and body worn measurements taken in a test site give false figures for calculating the true effective field sensitivity. To overcome this problem work is underway to design a new field test method which requires the construction of a Rayleigh distributed test site!

Whatever test methods are used the antenna has to have a low Q so that it is not adversely effected by the presence of the body. The r.f. front end noise figure of the receiver is normally in the order of 4dB to allow for an antenna gain of approximately -23dB relative to a dipole (for a typical 150MHz paging receiver).

The traditional receiver architectures for pagers are conventional double or single superheterodynes. Low power, one volt,circuits have been perfected with typically 10.7MHz or 21.4MHz first I.F. and 455KHz or 30KHz second I.F. Although front end crystal filters have been used and interest in SAW filters has been expressed, more recently revolutionary direct conversion architectures have been introduced [13] [14]. Such receivers use effectively no I.F. but directly convert the FSK frequency components of fc+ f, and fc- f into a digital output bit stream. This technique has the advantage that it can be almost fully integrated on a single chip and eliminates costly crystal filtering since all filtering is achieved at audio frequencies.

The digital decoders (or tone decoders) are generally fabricated in CMOS. Some manufacturers use available microprocessors, others custom LSI chips. The tone decoders in CMOS are almost exclusively supplied by a UK company, CML, although switched bipolar active filters are still extensively used. The techniques employed for obtaining bit synchronisation in digital decoders vary, some manufacturers use an asynchronous system in which continuous multi-sampling of the bit stream finds frame synchronisation without the need for a preamble comprising many data transitions. The POCSAG preamble comprises at least 576 bits of bit reversals to aid bit synchronisation and to contribute toward battery economy techniques. The power requirements, acquisition times and decoder complexity are all related to the bit/frame synchronisation techniques used. It is an area, together with battery economy techniques, over which many patent actions have been fought. The way in which pagers recover from long fades, and the percentage time that a receiver must be powered to guarantee maximum reliability are influenced by decoder performance in terms of data synchronisation as well as the inherent properties of the code format.

The POCSAG code uses batches and frames to give battery economy. The r.f. circuits which require most of the battery power (typically 5m watts) are switched by the decoder (typical power 200µ watts) in accordance with the code format. In the POCSAG case this is only during the synchronisation codewords (32 bits) and the one specified frame period (64 bits). However, the r.f. circuit and its interface characteristics determine what establishment time must be allowed before an acceptable bit stream can be presented to the decoder. This establishment time can have a major influence on battery life. During continuous transmission of codewords a typical POCSAG pager will be dissipating approximately 24% of its maximum receiver power and only 8.5% when receiving no r.f. carrier or a carriernot modulated with POCSAG codewords. Battery life is thus dependant not only on the pager on time but also the traffic loading of the paging network. Although high currents of 20mA are used for alerting or display illumination 'on' average duration of seconds per day, dissipates out neglible extra average power.

Bit error correction characteristics of the digital codes are used in different ways. If only one combination of, say 32 bits as in POCSAG is acceptable with perhaps, no more than 2 bit errors, simple correlation techniques can be employed, and even with a few acceptable combinations, many decoders use such correlation techniques on synchronisation and address data. For data message codewords in which no known combination of information bits can be predicted, two bit random error correction requires considerable computing power. In addition, the probability of falsely correcting such message data rises dramatically. The Motorola code uses eight interlaced codewords to give an effective increase in bit error correcting capability. However, burst error correction techniques can also be used, with good results on the POCSAG code.

Decoders are also required to store in memory many calls received, each of a different type, and to be able to generate specified and distinctive alerting patterns. A method of programming the individual pager addresses is a further requirement which is achieved by differing techniques, the most popular is the use of a fusible link diode or transistor array.

Message pagers require multi digit LCD devices and suitable driving circuits, though some pagers have used single or double digit LED indicators at the expense of battery life. The greater memory requirements of message pagers have resulted in an increase in functions required by the pager user. Radiation from logic circuits, ie. decoders, micro controllers, display drivers, into the R.F. front end via the loop antenna can desensitise the paging receiver, this problem must be carefully considered during design.

Message pagers have typically 16 character displays and can store many different messages with memory capacities greater than 500 characters. The ability to off load stored message data onto a printer is also available.

9.7 PROPAGATION

The standard techniques for predicting r.f. field strengths apply equally well to paging systems as to other radio systems. However, because of undefined antenna orientation, the problem of identifying building penetration attenuation and the relatively short duration of each address codeword, safety margins are required and also paging system performance is defined in terms of paging probability as a percentage of known locations in a coverage area. In any paging system coverage area there will invariably be some deep nulls, for example in lift shafts or underground stations, where the r.f. signal level is highly attenuated or shadowed. Another propagation hazard is in the construction of buildings with reflective glass windows which not only eliminate ultra violet and infra red radiation transmission into the building, but also r.f. radiation. In such cases, signal penetration may require the use of radiating cables driven from a roof-top antenna.

Many paging locations especially inside buildings are subject to multi path propagation and the r.f. field strength can then be shown to vary with position according to a Rayleigh distribution. This means that to achieve a 50% probability of successful paging for a static pager the pager sensitivity must equal the median level of r.f. field strength in the near vicinity of the pager location. To increase this probability of the pager being located in a position where the r.f. field will be equal to or greater than its sensitivity to say 90% and additional 9dB in the median level is required.

For much of its life a pager is stationary, however, it is subject to movement when the user is walking or travelling in a vehicle. When moving the receiver will pass through deep fades and the effect of these will be influenced by (a) the rate of movement, (b) length of message, (c) the median level of the r.f. field, (d) the bit rate of the code, (e) the error correction algorithms employed. Because of the trade offs already identified between sensitivity, falsing and information throughput each code has its supporters and responds differently in the moving environment. Table V gives some sensitivity comparisons which are typical for either POCSAG or GSC ([15], [7]). Because of the interlaced code used in the GSC it is claimed to be more sensitive in the long message moving environment and thus could save transmitters. However, because of a lower capacity more r.f. channels and thus transmitters would be required to serve the same number of subscribers.

Following extensive testing in London of POCSAG message pager [15], Tridgell and Denman proposed the following equation to calculate the additional field strength required over that needed for tone only paging for 80 character Alpha message paging:

$$\text{Message Loss (dB)} = 3.2 + 0.055W$$

where W is the percentage of time spent walking, over the range 0-50%.

From an analysis of typical behaviour patterns it is expected that between 15 to 30% of a day is spent walking. The loss of sensitivity at higher speeds is not seen to be a problem, for these speeds will only be achieved in the street and thus not suffer the same loss due to building penetration which must be used to guarantee adequate coverage within buildings, see Table VI. The penetration loss of a motor vehicle is in the order of 8dB and the penetration loss within a building will vary with building type and floor.

	Static	3mph	30mph
Tone only	0dB	1dB	2dB
Numeric 10 digit	1dB	3dB	3dB(5)
Alpha 40 character	2dB	7dB	5dB
Alpha 80 character	3dB	10dB(17)	7dB(13)

TABLE V. Relative median r.f. signal levels for 90% success rate. 150MHz

The figures in brackets in Table V are computer simulated results for POCSAG with only 1 bit error correction [7] but were not found in field trials [15].

It can be seen that higher field strengths are required for successfully receiving 80 character messages for the common condition of walking inside buildings.

Wide area paging systems, sometimes called WAPS, by definition must provide a very large coverage area. To achieve this, within the constraints of narrow band mobile radio, many fixed transmitters must be used. The maximum effective radiated power normally permitted in Europe is 100 watts. Although in some cases much higher powers are allowed eg. in remote districts or near broadcast frequency bands.

Quasi synchronous transmission is used to achieve the required area coverage and the characteristics of some of the paging codes reflect this. For example the POCSAG bit rate is 512 bits per second which is also its baud rate thusallowing differential delays in the transmission paths to each transmitter modulator of 250μsecs. This is calculated in the following way by assuming a maximum allowable confusion period [5] of 25% of a signalling element.

25% of signalling element	= 488.3µS
10% allowance for telegraphic distortion (±5% isochronous)	= 195.3µS
Tx rise time tolerance	= 25µS
Radio propagation delay differential (10Km path difference)	= 30µS
Therefore, max. permissible line differential delay	= 238µS

The choice of offset frequencies for paging transmitters radiating simultaneously has caused some controversy over recent years. The POCSAG report identified offsets of 500Hz and 1000Hz because it was influenced by a Japanese pager [5] which shows that offsets should be at least twice the signal fundamental frequency. The POCSAG report was followed by Telecom Australia and the CCIR Rec 584 in this aspect but fortunately not by British Telecom! Because lat er work shows that in the real world the cancellation and enhancement caused by the two (or more) r.f. signals adds another Rayleigh distributed field to that which already exists, and offsets have little effect on signalling probability, but do have a marked effect on receiver design problems, particularly as the channel separations are reduced. So offsets of ±50Hz (max. ±200Hz) are normal, and preferred by Rx designers.

To cover a country the size of the UK with 100 watt transmitters about 500 sites are required. As it is not necessary to radiate all calls in all areas, but, for system capacity reasons, desireable to radiate different calls simultaneously in different areas, a mixture of sequential and simultaneous transmission is often employed. A zone structure is used so that two zones transmitting together do not overlap but are separated by zones that are transmitting in a different time slot. Major zones and minor zones can be identified to relate to population densities and more time allocated to major zones. Figure 6 illustrates a commonly used arrangement.

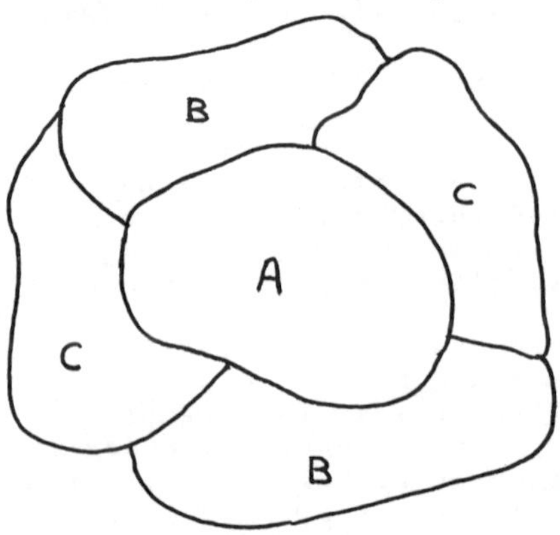

Fig. 6 :

A zone structure as shown in Figure 6 does add more complexity to the propagation models used and can result in a need for more sites and thus cost to overcome co channel interference and re-use distance problems. The Euro signal system uses a zone structure but, unlike the UK system, separates zones by frequency and not time. This has the advantage of not reducing the calling rate (which, for Eurosignal is very low), but has the disadvantage of increasing the complexity of the pager. There are three zones in Germany, using two frequencies, and seven zones in France, using four frequencies. In Germany approximately 45 transmitter sites are used, radiating at 87.340MHz and 87.365MHz, most with approximately 100W ERP, but eight use 2KW, and two at Hamburg and Stuttgart, radiate 156KW . All are amplitude modulated except the two 156KW transmitters which are frequency modulated.

The choice of frequency for the ideal paging system is difficult. Propagation losses increase with frequency, but building penetration decreases, as does man made noise.

Low frequencies, below 50MHz, have co-channel interference problems and channel re-use difficulties. Receiver noise factor increases with frequency and, for given antenna gain, leads to less sensitivity and/or more costly pagers at the higher frequencies. For these reasons where the spectrum has been available the 100 to 200MHz band has been chosen for paging.

Some building penetration measurements have been made at different frequencies in Japan and the UK. The lower figures are those made in the UK and shown in Table VI.

Frequency	150MHz	250MHz	400MHz	800MHz
Building Penetration Attenuation	14-22dB	18dB	12-18dB	17dB

TABLE VI. Building Penetration

Subsidary Communications Authorities (SCA's) paging has been permitted in the USA since May 1983 and similar broadcast service type paging is allowed in other countries. The paging service makes use of sub channels of an FM broadcast transmitter. In the USA the FCC has changed its rules so that (a) sub channel operation is permitted whether or not the main channel is in operation and (b) the FM broadcast is expanded from 75KHz to 99KHz. Sub carriers of 67KHz or 92KHz are used and occupy 15KHz each. Any of the paging signalling formats can be used. Although broadcast transmitters operate at high powers some even at 100KW with 2000 feet antenna heights the power in the sub channels is only a portion of the total ERP.

The coverage is designed to service radio receivers with better antenna systems than those found on pagers but nevertheless this approach to paging can also be successfully used.

9.8 PAGING TRANSMITTERS

Transmitters for paging systems are greatly influenced by the system modulation, the power relating and system switching requirements. However, for most new digital paging systems, like POCSAG, the low frequency components in the modulation mean that the conventional PMR phase modulated transmitters are unsuitable, and true FM or FSK modulators are required. Continuous rating and comprehensive monitoring and control are also important elements as reliability and remote operation are essential. Because of the modulation limitations it has restricted the rapid introduction of low cost POCSAG pagers to already existing tone signalling systems.

The transmitter frequency stability requirements are, for a quasi synchronous system, in the order of $\pm 2 \times 10^{8}$ (from $-10°C$ to $+50°C$) with a drift of $\pm 1 \times 10^{-7}$ per month. The offset frequencies are not considered as important for straight digital alert only paging as they were, but to ease the design and maximise the performance of the many hundreds and thousands of pagers, the carrier stability is important. A high stability at the Tx allows an increased drift in the pager which avoids servicing pagers and also reduces maintenance adjustments to the transmitters.

Two channel transmitters are used by BT which because time division zoning techniques are used, the transmitter can be switched between channels rather than to off. Occupying 50% of the time on channel 1 and 50% on channel 2 in the major traffic zones.

9.9 SYSTEM ARCHITECTURES

In the USA paging is a major business and has grown by essentially private operators building local city wide systems, which operate to the most part independent of other cities and their systems. Frequencies, signalling systems and number allocations are all different. Two tone, five tone, three or four different digital systems, display and voice are all mixed. In smaller European countries the user would like to move from city to city and still use his pager. Thus more integrated systems have been conceived from the outset. These systems either have a single frequency allocated for paging in all cities or a multi-channel pager is required. The problems of frequency allocation, and co-operation between independent operators have been inhibiting factors for National Paging in the USA, but new moves to allocate over 40 channels nationally in the 930MHz band will overcome this problem.

A transmitter network must be established and some System Architectures have been conceived around the availability of an already existing broadcasting network. Notably the Swedish MBS system which uses a pilot tone on the national FM sound broadcasting service. A new paging service in the USA is also proposed using the sound broadcast transmitters. The Swedish system is national and so the multi channel pagers used are costly. Although at first this seems an ideal solution to the nationwide transmission problem, there are some more drawbacks; the broadcasters are thinking of new ways of using their outer frequency bands on their transmissions notably for sending data on programme content, and the coverage offered is not as complete as it needs to be for paging. This is because the broadcast receiver has a better antenna in general and the listener knows if he is in range because he is listening and the effective power radiated in the narrow band pilot tone carrier is only a fraction of the very high broad band station power.

The other need for multi-channel pagers is as mentioned before, the weakness of the signalling system (eg. Eurosignal). Even if a single frequency is available and the calling rate possible is very large it is still desireable not to send calls out in all regions unnecessarily. This can be achieved by not allowing paging in country areas, thus creating uncovered zones between cities. In a large country like the USA this is acceptable but in a small country like the UK it is not. The compromise is to reduce the effective calling rate by temporarily switching off the transmissions in the large cities (ie. areas of high traffic density) and for these time slots transmit in the separating country areas. The BT system has chosen a three phased zone structure as illustrated in Figure 8 for a total of 40 zones. Zone type A, the big city zone type, is enabled for 1 minute every 2 minutes. The number of pagers that can be supplied on one radio channel depends on (a) the traffic density during a peak busy period, (b) the calling rate of the code, (c) how the pager population is distributed between zones in a zoned system and (d) code address capacity. Table VII compares some codes and gives maximum pager populations for a busy hour traffic density of 0.1 calls per subscriber.

SYSTEM	ADDRESS CAPACITY	CALL RATE PER SECOND (TONE ONLY)	MAX. PAGER POPULATION FOR TRAFFIC = 0.1 CALLS/SUB/BUSY HOUR					
			TONE ONLY		10CH NUMERIC		40CH ALPHA	
			NONE-ZONED	A-ZONE	NONE-ZONED	A-ZONE	NONE-ZONED	A-ZONE
EIA (2 tone)	3.5×10^3	0.21	3500	3500	-	-	-	-
EIA (5 tone)	1×10^5	4.70	100000	84600	-	-	-	-
Euro Signal	7.5×10^6	1.20	43200	21600	-	-	-	-
NTT	4×10^6	5.50	198000	99000	65880	32940	12240	6120
GSC	4×10^6	3.30	118800	59400	59400	29700	19800	9900
POCSAG	8×10^6	15.06	542160	266562	180720	88854	36000	17700

TABLE VII.

The use of sequential zone transmissions is good if most users request regional coverage and can accept a possible delay of say 2 minutes before transmission of a page. If the sequence period is reduced then the effective calling rate is reduced by the need to transmit at least 3 preamble batches (1 second each). The ratio of the major A zone time to the minor B and C zone times could be dynamically adjusted to optimise the system performance especially if everyone wants just one major zone (eg. London). Other systems allow the user to dial selected zones so that calls are only transmitted in zones where the pager is expected to be. This is good for traffic optimisation, and good for the user because he can be called anywhere in the whole country without paying for coverage in every zone in advance for every day of the year.

9.10 PAGING TERMINAL CONCEPT

A paging terminal is essentially a computing system that interfaces the telephone network to a network of r.f. transmitters. It has to perform a number of important functions.

(a) Sufficient input trunk ports must be provided to offer the acceptable grade of service for the estimated traffic.

(b) A file for each subscribers telephone number. This file must contain a complete register of information relating to the number dialled. The following data will be stored:

Subscriber number (telephone number)
Valid (Yes/No)
Group Number (Yes/No)
Pager Code (or codes) to be transmitted (RIC + function code)
Group size
Pager type
Zones to be paged
Radio channel
Priority level
Message type (tone/numeric/Alpha)
Call forward (transfer)
Page present in system (Yes/No)
Access Voice bank
Customer Number
Account Number

(c) Access to each subscriber file for updating in the event of failure to pay, change of pager due to fault, or re-assignment by system operator, and change of availability (present/not present in system), or transfer instructions by pager user.

(d) Temporary registers to allow queuing of calls prior to transmission. This allows sequential zoning to be served and a choice of signalling systems to be permitted. It also accommodates very high peaks in the traffic demand.

(e) The storage of voice messages, allows for either pre-recorded announcements common to all subscribers (eg. "paging call accepted", "number invalid", "pager not present" etc.) or user defined. Such voice storage and retrieval by pager users requires large memory capacity but adds considerably to the usefulness. Calling subscribers can be given information on the availability of the pager user with personally recorded voice messages and invited to leave personal messages only recoverably by the pager owners with security codes.

(f) The sequencing of transmissions in a zone structure and the encoding of data into a form suitable for modulating the radio transmitters. The encoding of the modulating signals can be achieved in district controllers which are remotely sited and thus economic on line costs and allow the delay equalisation of lines to the transmitters to be preset and simply maintained.

(g) The terminal must have a queuing capacity to group calls in registers relating to frame numbers if POCSAG or similar signalling systems are employed or if multi pager types are to be serviced. The formation of message codewords must also be achieved and a temporary memory large enough to handle the traffic provided.

(h) Redundant operation of main and standby equipment is normally necessary to provide the level of system reliability required. Alarm monitoring of remote transmitter equipment is also necessary.

(i) Address capacity and networking is also important. To enable system growth and optimise on telephone network interfacing the distribution of call processing and transmitter control is a very useful property.

(j) Traffic analysis and possibly billing and call charging is often requested. The ability to record calls enables traffic peaks and the number of calls for each subscriber to be identified. The generation of additional meter pulses is another way of charging for paging calls.

Terminals which handle many different types of pager signalling formats are often called SWAP terminals. SWAP is an acronym for System Wide Area Paging. This was first conceived in Canada to overcome the problem of single source supply and was adopted by British Telecom and other operators in the USA and Australia. However, it does not offer a good system concept because of the time wasted by the transmission of many different preambles and the incomplete batching. The use of voice paging is attactive to the user but not the system operator unless he can obtain enough channels to support the user demand or if the charges can be continually increased without objection. Techniques of voice storage and compression are used to aid this process.

9.11 THE FUTURE

The widespread use of paging for the entire population is expected to follow a similar pattern to the development of the telephone. By 1990 1:17 of the population of the USA is expected to own a pager, which by extrapolation could mean that approximately 1:25 of the population of Europe could possess a pager also. A move to higher frequencies and higher signalling speeds may prove necessary to support demand for more comprehensive message pagers with large memories, multi-line alpha displays, voice synthesis and hard copy printers. The extensive use of computers to supply data to selected users over the paging network and the provision of voice recognition in the paging terminals will all speed the appreciation and spread of pagers. The falling prices and dimensions of the bodyworn receiver could finally put the pager in the watch and calculator category. Not being accessible when out of the office or away from home will no longer be acceptable.

REFERENCES

1. British Telecom specification RC4501D for Radiopaging Receivers, March 31st 1980, (see also RC4511 and RC4512 for Numeric Message Display and Alpha Numeric Message Display, April 1982).

2. Telecom Australia. Specification 1343 Issue 2 Radio Paging Receiver (POCSAG Digital Code) October 1981.

3. I.E.C. Publication 489 Methods of Measurement for Radio Equipment used in the Mobile Services. Part XX Selective Calling. Receiver Decoder Equipment R.F. Receiver Decoder Parameter Measurements 12F (Central Office) '75. April 1983.

4. KOMURA, M., HAGIHIRA, T. and OGASAWARA, M., (November 19877) New Radio Paging System and its Propagation Characteristics. IEEE Trans. Vehic. Tech. Vol. VT-26.4, 362-366.

5. KOMURA, M., etal "New Radio Paging System" Japan Telecom Review July 1977 Vol. 19,3

6. British Telecom. A Standard Code for Radiopaging. Report of the Post Office Code Standardisation Advisory Group (POCSAG). June 1978 and November 1980

7. SANDVOS, J.L., "A Comparison of Binary Paging Codes" IEEE Vehic. Tech. Conference May 1982.

8. TRIDGELL, R.H. "The CCIR Radio Paging Code No. I A New World Standard". TEE Vehic. Conference May 1982.

9. MAKITALO, O., Fremin G. "New System for Radio Paging over the FM Broadcasting Network" TELE (English Edition) Vol. XXII No. 2 1970.

10. MABEY, P.J., "Digital Signalling for Radio Paging" IEEE Trans. Veh. Tech. Vol.vt-30th May 1981.

11. CEPT. Recommendation T/R6, Systeme Eurosignal.

12. SHARPE, K. "The Eurosignal Radio Paging Systems" T.C.S. August 1975.

13. VANCE, I.A.W., "Fully Integrated Radio Paging Receiver" IEE Proc F. Vol. 129 1, 2-6 1982

14. FRENCH, R.C., "A High Technology VHF Radio Paging Receiver" IEE Conf MRST Sept. 1984 Conf Proc Page 11.

15. TRIDGELL, R.H., DENMAN, D. "Experience of CCIR Radiopaging Code No. 1 (POCSAG) for Message Display Paging" IEE Conf. Mobile Radio Systems and Techniques Sept. 1984 Conf. Proc. Page 194.

Chapter 10

Regulatory control of land mobile radio

R.M. Skiffins

10.1 INTRODUCTION

The steady rise in fuel prices combined with a need for improving the efficiency of industry and commerce have resulted in an increased demand for new and more advanced mobile radio systems. This demand has been intensified by the increasing general awareness in the fields of communications and information technology. Thus there is a growing requirement for the transmission, processing and storage of information which is affecting all areas of business, commerce, government, education, health and transport. This information technology revolution is creating a demand for advanced communications networks which allow access to this information whenever and wherever it is needed.

Land mobile radio systems provide an important link in this process of information transfer, particularly with the increasing integration of line and radiocommunications networks. However the shortage of suitable radio-frequency spectrum to meet this growing demand is one of the major problems facing radio regulatory authorities throughout the World.

In the United Kingdom, the land mobile radio service is expanding at a rate of 8-10% per annum, and demand is expected to grow to the end of the century. At present there are some 17,000 different users having in excess of 20,000 base stations and more than 300,000 mobiles. By 1990 the number of mobiles is expected to exceed 700,000 and could well exceed 2 million by the end of the century.

Over the years, the land mobile service has been in the forefront of technological change, accepting progressively narrower channel spacing and introducing increasingly sophisticated equipment in order to maintain a usable grade of service against this background of consistent and rapid growth. However, despite its importance, in the frequency bands between 30 MHz and 1000 MHz, the most useful part of the frequency spectrum for mobile radio, it occupies only some 30 MHz or 3% of the available spectrum.

Careful management of the frequency spectrum is necessary therefore to ensure efficient and equitable utilisation in the face of many competing and expanding demands, and to ensure freedom from harmful interference.

The process of spectrum management starts in the international arena, where agreements are reached on the broad categories of use to which the various radio frequency bands may be put. Through the national frequency planning machinery , these sub-divisions of the spectrum are allocated to meet national requirements as, for example, between civil and military uses.

The bands thus available for civil land mobile radio are then subject to detailed planning and frequency assignment in order to ensure the maximum possible utilisation within the United Kingdom.

International spectrum management is accomplished through the media of the International Telecommunications Union (ITU), the Conference of European Postal and Telecommunications Administrations (CEPT) and the International Electrotechnical Commission (IEC). Through the levels of mutual understanding and acceptance reached in CCIR, CEPT and IEC, the UK mobile radio industry is more readily able to compete in overseas markets, trade barriers are eased and regulatory equipment type acceptance procedures are simplified.

10.2 THE INTERNATIONAL TELECOMMUNICATIONS UNION (ITU)

The ITU is the specialised agency of the United Nations recognised as competent in the field of international telecommunications and comprising more than 150 member nations. The basic instrument of the ITU is the International Telecommunications Convention and associated Administrative Regulations which rank as a treaty, the provisions of which are binding on Governments which have signed and ratified or acceded to it.

The purposes of the ITU as set out in the Convention are:

- to maintain and extend international co-operation for the improvement and rational use of telecommunications of all kinds.
- to promote the development of technical facilities and their most efficient operation.
- to harmonise the actions of nations in the attainment of these ends.
- to effect the allocation of the radio frequency spectrum and the registration of frequency assignments.
- to coordinate efforts to eliminate harmful interference and to improve the use made of the frequency spectrum.
- to foster the establishment of rates at levels as low as possible consistent with the provision of efficient service and the necessity for maintaining independent financial administration.
- to foster the creation, development and improvement of telecommunications networks in new or developing countries by all possible means, especially by United Nations programmes.
- to promote the adoption of measures for assuring safety of life through telecommunications services.
- to undertake studies, formulate recommendations and to collect and circulate information for the benefit of all members of the Union.

For the allocation of frequencies, the World is divided into three regions; Region 1 covering broadly Europe, the Middle East, Africa and the USSR, Region 2 - the Americas, Canada and Greenland, and Region 3 - the Far East, Australia and New Zealand.

The work of the Union is reviewed and the Convention revised as necessary by a Plenipotentiary Conference held about every six years, the last being in October 1982.

Acting on behalf of the Plenipotentiary Conference, the Administrative Council forms the governing body of the Union, facilitating the implementation of the Convention and Regulations and of the decisions of conferences.

The Administrative Regulations are kept up to date by World Administrative Conferences. A World Administrative Radio Conference may encompass all aspects of radiocommunications as did the 1979 WARC or, more usually, will limit its scope to one particular service. A WARC for the mobile services will be held in 1987.

One of the most important articles of the Radio Regulations is the International Table of Frequency Allocations. Covering all three regions of the World, the Table provides allocations of use for all frequency bands between 9 kHz and 275 GHz. An example from the international table of frequency allocations is given in Figure 10.1. In each region a given frequency band may be allocated exclusively to one service or shared between two or more services. In the latter case, the relative status of each service is defined as primary, permitted or secondary. Primary and permitted services have the same rights except that in the preparation of frequency plans, primary services have the first choice of frequencies. Secondary services must not cause harmful interference to primary or permitted services, and cannot claim protection from harmful interference from such services.

It has to be recognised therefore that this international framework of control imposes constraints on the way individual countries can use the spectrum, but that these constraints have the objective of securing, for the benefit of all, the maximum, effective, non-interfering usage of a finite global natural resource.

The Union is supported by four permanent organs:
<u>The General Secretariat</u> is responsible for carrying out the directions of the Plenipotentiary Conference and of the Administrative Council, for the administration and financial services of the Union and for the custody and maintenance of records and archives.

<u>The International Frequency Registration Board</u> which, under the control of five board members, is responsible inter-alia for the orderly recording of frequency assignments, for advice concerning improved utilisation of the radio spectrum and technical assistance in the preparation and organisation of Administrative Radio Conferences.

<u>The International Consultative Committees (CCIs)</u> The study of technical and operational matters and questions concerning the establishment, development and improvement of telecommunications in new or developing countries are the principal functions of the International Telegraph and Telephone and the International Radio Consultative Committees (CCITT and CCIR). The work of the CCIs results in recommendations which while not binding are generally observed by member nations. Each consultative committee works through:

- a plenary assembly, meeting normally every three years
- study groups set up to deal with questions remitted to them for study.
- a Director supported by a technical and administrative Secretariat.
- laboratories or technical installations set up by the Union.

The work of the CCIR is divided among some twelve study groups addressing the problems of various radio services as well as general matters such as propagation, spectrum utilisation and monitoring.

Study Group 8 is concerned with Mobile Services including:
- Land Mobile Service
- Maritime Mobile Service
- Future Global Maritime Distress and Safety System
- Maritime Mobile Satellite Service
- Land Mobile Satellite Service
- Aeronautical Mobile Service
- Amateur Service

For the land mobile service, CCIR texts include:
- Technical characteristics of equipment
- Protection ratios and required minimum field strengths
- Radio paging systems
- Trunked mobile radio systems
- Single sideband modulation
- Digital transmission
- Public mobile telephone systems

The CCIR is also responsible for providing the technical bases for Administrative Radio Conferences.

The CCIR provides the primary technical mechanism for achieving harmonisation in the development and co-ordination of radiocommunications systems on a worldwide basis.

10.3 CEPT

The Conférence Européene des Administrations des Postes et des Télécommunications was founded in 1959 as an official, non political, recommendatory body for the development and improvement of postal and telecommunication co-operation between the countries of Europe. Comprising 26 member nations, CEPT carries out its work on the technical, operational and administrative fields through the joint study of common problems. The conclusions of these studies take the form of recommendations which member administrations apply where possible according to their particular circumstances. Membership is open to any European postal and telecommunications administsration which is a member of the Universal Postal Union (UPU) or the ITU, although in fact, of the Eastern European countries, only Yugoslavia is a member.

The organisation works through two main Commissions, the 'Postal' Commission and the 'Telecommunications' Commission, each of which have a series of working groups. The working language of CEPT is French and all documents are produced in that language.

Reporting to the Telecommunications Commission are ten working groups:

T/GT1 Telegraph
T/GT2 Telephone
T/GT3 Radiocommunications
T/GT4 Television and Sound Transmission
T/GT5 General Tariffs
T/GT7 Services and Facilities
T/GT9 Long Term Studies
T/GT10 Data Communications
T/GT11 Switching and Signalling
T/GT12 Transmission

Under the Radiocommunications Working Group, seven technical committes are constituted of which four are of particular interest for land mobile services:

- R19 Equipment for the land mobile service
- R21 Radio-frequency management
- R22 Harmonisation of low power devices
- R31 Compatibility between land mobile and broadcasting services

A further group known as GSM (Groupe Special Mobile) has been created in order to formulate the requirements for a 900MHz pan-European cellular mobile radio telephone system for the 1990's.

The recommendations of CEPT include such subjects as pmr systems, data and signalling systems, paging, telemetry, citizens band, frequency planning and co-ordination in border areas.

10.4 INTERNATIONAL ELECTROTECHNICAL COMMISSION

Under the auspices of the International Standardisation Organisation (ISO) the IEC seeks to establish standards on all kinds of domestic and commercial electrical and electronic equipment.

Radiocommunications matters are handled by Technical Committee 12 of IEC and mobile radio matters by Sub-Committee 12F. The resulting IEC Standards, which are adopted in the UK as standards of the British Standards Institute (BSI), are widely used as the basis for national and international mobile radio equipment regulatory provisions.

10.5 THE NATIONAL REGULATORY FRAMEWORK

Within the United Kingdom, under the provisions of the Wireless Telegraphy Act 1949, control of the civil use of radio is vested in the Secretary of State for Trade and Industry who is advised on radio regulatory matters by the Radio Regulatory Division. This division comprises two administrative branches which are responsible for frequency policy, regulatory and licensing matters and the Directorate of Radio Technology comprising some 130 professional and technical engineers which provides the necessary technical advice. Within the constraints imposed by the International Radio Regulations, the United Kingdom, like other countries,is free to allocate and sub-divide frequency bands for different uses as it sees fit. The interests of users are co-ordinated within a framework of interdepartmental committees which at the higher level formulate the overall policy concerning telecommunications and radio matters and the broad allocation of radio frequencies.

At a lower level, detailed frequency planning is carried out in order to produce and update the UK national frequency allocation table and a national register of frequency assignments. The interdepartmental machinery provides also for the preparation and coordination of the UK input to Administrative Radio Conferences of the ITU.

In a move towards greater openness in Government and following the recommendations of Merriman (8), the United Kingdom frequency allocation table will be published in 1985.

Associated with this frequency management machinery is a further organisation for the clearance of radio sites, the aims of which are to foresee some of the special problems that can arise where transmitters and receivers operate in close proximity and devise remedial measures before sites are taken into use.

On matters affecting the use of mobile radio, the Secretary of State is advised by a standing body known as the Mobile Radio Committee on which manufacturers and users are represented, and in addition, a continuous dialogue with the mobile radio industry is maintained through the Land Mobile Working Party.

National regulation of mobile radio necessarily extends far beyond the allocation and assignment of frequencies. Since the extent to which frequencies may be re-used is dependent, in part, on the technical characteristics of the equipment used, performance specifications for land mobile equipment are prepared by the Directorate of Radio Technology in consultation with industry and users. Adherence to these standards constitutes a major condition for the issue of a mobile radio licence under the Wireless Telegraphy Act. Standardisation of equipment performance characteristics facilitates planning of mobile services in order to avoid harmful interference between systems and to make optimum use of the frequency bands.

It is thus the executive function of the Radio Regulatory Division to process applications for private mobile radio users, assign suitable radio frequencies, control standards of equipment used and issue the appropriate licences.

With the rapid advances in technology however, and the increasing competition for spectrum it is becoming increasingly necessary to make valued judgements of competing applications. These aspects of competition in the supply of telecommunications services will in the future also be of interest to the new Office of Telecommunications (OFTEL) under the Director General of Telecommunications.

10.6 COMPARISON WITH THE USA

It is interesting here to compare the national regulatory infrastructure in the UK with that which exists in the USA.

Under the Communications Act 1934, overall control of the frequency spectrum is vested jointly in the Federal Communications Commission (FCC) and the National Telecommunications and Information Administration (NTIA). Broadly NTIA is responsible for all federal government use whilst the FCC controls all other uses.

The FCC is an independent body yet accountable to Congress. Its responsibilities include those of general frequency policy, allocation and providing technical and engineering advice.

Considerable openness is adopted in the frequency management process, public consultation being achieved through Notices of Enquiry, Notices of Proposed Rulemaking and occasionally public hearings.

The work of NTIA on the other hand, is undertaken largely through the Interdepartment Radio Advisory Committee (IRAC) comprising representatives of those federal departments making greatest use of radio.

The FCC and NTIA jointly undertake long-term spectrum planning and a national table of frequency allocations is published by NTIA. Whilst the FCC specifies the technical characteristics of services and issues the appropriate licences, the assignment of frequencies to private land mobile radio users is generally undertaken by geographically located frequency co-ordinators having detailed local knowledge of frequency usage.

Under the Administrative Procedure Act, the FCC is required to make public all rules, orders, opinions records and proceedings.

10.7 SPECTRUM AVAILABILITY AND UTILISATION

The main task of the World Administrative Radio Conference 1979 was to carry out a detailed revision of the International Table of Frequency Allocations contained in Article 8 of the Radio Regulations. As such, the conference set the stage for radio spectrum utilisation until the end of the century.

For land mobile radio, certain decisions of the conference are of particular importance.

- The band 41-47 MHz was allocated to fixed and mobile services on a primary basis and will be relinquished by broadcasting services in the United Kingdom by 1 January 1985.

- Although the bands 47-68 MHz and 174-223 MHz remain allocated to broadcasting on a primary basis, the land mobile service was added on a permitted basis and will be available in the UK from 1 January 1985 following the recommendations of the Merriman Review (1). However the continued use for broadcasting in France and the Republic of Ireland will limit the effective use that can be made of these bands in many parts of the UK.

- The band 862-960 MHz was allocated to fixed and mobile on a primary basis and land mobile services in the UK have been given access to the band since 1 January 1982 on an equal primary basis with the fixed services.

- The band 100-108 MHz was allocated to broadcasting on a primary basis. Existing police and fire service communications in the band 97.6 to 102.1 MHz must be reallocated by 31 December 1989. In the band 104-108 MHz, mobile radio services in the UK remain protected internationally until 31 December 1995 although it may be necessary to re-allocate these before 1995 owing to national broadcasting requirements.

10.7.1 Bands I and III

In May 1984, the Government issued a consultative document on the future use of television bands I and III, (7) requesting comments on a number of key issues.

In particular, emphasis is placed on the concept of service providers, following the recommendations of Merriman. (8) Extensive use of trunked common base stations is advocated in order to achieve efficient spectrum utilisation in Band III whilst also achieving a higher grade of service.

Band III will present tremendous opportunities for expansion of the land mobile services, although, since the band will be shared with foreign television broadcasting, new spectrum management problems will be imposed, and certain parts of the band will not be available in particular areas of the country. Nevertheless there will be some 1000 12.5 kHz dual frequency channels in London, more than doubling the present number.

Band III will present the opportunity to exploit new technology not only through development of trunked systems, but also by the introduction of new modulation methods and more efficient communications techniques.

The restrictions on the use of frequencies in some parts of the country mean that users requiring a wide area service will require equipment which can switch across a range of channels.

A new equipment specification has been developed for Band III (9) requiring the use of synthesised equipment having a minimum of 5 MHz switching range, and it is hoped to employ a standardised digital signalling system based on 1200 baud FFSK.

Band III will provide also for the broadcasting ancillary services, as well as offering possibilities for paging and further expansion of radiotelephone services.

In addition, those areas of Band III that may not be used by conventional mobile radio systems, will have a ready application for low power services possibly using leaky feeder technology.

10.8 SPECTRUM REQUIREMENTS

In 1982 the Mobile Radio Committee reported the results of a study carried out by the Radio Regulatory Division to determine the spectrum requirements of the land mobile service to the end of the century. The resulting report (2) concluded that in order to achieve an overall channel loading acceptable to users, an additional 141 channels were required at that time and by 1990 a total of 1784 channels will be needed, requiring some 27 MHz of additional frequency spectrum. In addition, spectrum would be required for public radiotelephone systems, cordless telephones and requirements ancillary to broadcasting.The study was confined to an area covered by a 50 km radius of central London on the basis that meeting the needs of this area should automatically provide for a national solution. A questionnaire was issued to users in order to determine a pattern of use for each of the following categories:

- Central Government
- Local Government
- Manufacturing/Commercial
- Services
- Public Transport
- Other Transport
- Utilities (excluding the Post Office)

Projections of growth made from this exercise were compared with similar projections made by the Home Office in 1975 (3) and with the actual usage achieved by 1980. The results are summarised in Table 10.1.

In order to assess the future spectrum requirements, an agreed channel loading criteria was required which, in the mean busy hour, would provide the required grade of service.

Standard telephone traffic theory (Erlang C) was used in the analysis, which although not providing an adequate representation of a single p.m.r. shared channel, was shown to give a good estimate of global channel requirements.

From Erlang C, the normalised mean waiting time

$$\frac{W}{H} = \frac{A}{1-A}$$ and

the probability that the waiting time exceeds t secs

$$P\ (D>t) = A\ \exp-\left[\frac{(1-A)t}{H}\right]$$

where W = mean waiting time in the mean busy hour
H = mean holding time
A = offered traffic per channel (Erlangs)

From the data collected, the mean holding time and mean waiting time acceptable to the user was used to determine the normalised mean waiting time and the channel loading for each user category. The number of mobiles per channel to give the required grade of service during the mean busy hour was determined and, knowing the proportion of the total mobiles using the channel during the busy hour, the number of licensed mobiles per channel to give an acceptable mean waiting time for each user category was established. The results are summarized in Table 10.2. From this loading criteria, and given the total number of mobiles in each category, the total number of channels required was established.

10.8.1 Cellular Radio

Public mobile telephone systems employing small cell techniques will account for probably the largest single new application of mobile radio in the foreseeable future. An initial allocation of about 7.5 + 7.5 MHz has been made in the 900 MHz band in which two competing service providers have been licensed to provide a service starting no later than 31 March 1985, and covering 90% of the population by 1989.

The system known as TACS (Total Access Communication System) is a derivative of the American AMPS system, modified to operate in 25 kHz spaced channels.

Initial predictions for TACS are in the order of 10 -25,000 mobiles in the first year, with up to 10,000 base stations being required to give national coverage.

Further growth of TACS will be accommodated in an additional 10 + 10 MHz at 900 MHz which will become available towards the end of the 1980s.

Within Europe this band has been earmarked for a pan-European mobile telephone system known as GSM (after the Groupe Special Mobile of CEPT in which the discussion of this system is being conducted). Whether these discussions will lead to a truly European solution which at the same time could be adopted in the United Kingdom is, at this point in time, difficult to predict.

10.8.2 Cordless Telephones

Introduced in the UK in 1982 on eight frequency channels between 47.45 and 47.55 MHz paired with 1.632 to 1.792 MHz, the cordless telephone is one of a number of emerging land mobile applications with potential for the high volume domestic market.

Some 77% of Britain's 20 million households have a telephone with 14% of these currently having an extension telephone. There is a growing market for extension telephones, and cordless telephones will have a part to play in this.

Recent studies indicate that a penetration of 7-10% of all telephones should be allowed for.

In the short term, an improved first generation cordless telephone operating in Band I would provide for considerable expansion.

In the medium term, a second generation cordless telephone system operating in the 900 MHz band could need to handle upwards of 1000 cordless telephones per sq km in a typical suburban environment.

In the conurbations however, the possibility of an all wireless PBX (using cordless extensions to provide total flexibility in the planning of office premises) could raise the penetration figure by at least one order.

10.8.3 Low Power Telemetry and Telecontrol

Another area of potentially high volume penetration, low power telemetry and telecontrol systems are used in a wide range of applications.

Alarm systems for the elderly and infirm, domestic security (burglar and fire alarm systems), commercial security (eg identification tags on retail goods), stock control, medical telemetry and industrial telemetry and telecontrol all figure in this diverse and expanding market.

In 1984, the Government issued a consultative document (10) outlining considerations on the future control of these devices.

The probability of some deregulation in this area, through the exemption from licensing, could have a significant impact on the growth and spectrum requirements here.

10.9 IMPROVING THE UTILISATION OF MOBILE RADIO CHANNELS

As the private mobile radio bands have become more congested, users have been forced to share channels particularly in the major conurbations. Channel sharing causes contention difficulties and inefficiency arising from the random nature of mobiles accessing the channel causing lost calls and increased waiting times. Although providing no improvement in channel utilisation, the subjective effects of this congestion can be reduced by the use of selective signalling techniques.

Common base stations, trunking systems and the use of data transmission, together with improved techniques for the assignment of frequencies, provide for significant improvements in spectrum efficiency.

10.9.1 Common Base Stations

In major conurbations, it is common for several quite separate organisations to operate on the same frequency each having a base station in close geographical proximity to the other. The use of the channel is undisciplined since no co-ordination exists between one user and another. Common base station (cbs) operation provides the possibility to adopt a co-ordinated use of the channel, giving users a better grade of service by eliminating the 'free-for-all' approach of co-channel sharers. The reduction in the number of base stations and antenna structures, reduces radio interference, improves the environment and eases the problems of frequency assignment since the frequency re-use distance is more clearly defined than is the case for a number of individual base stations. Selective signalling is used to ensure a measure of privacy for users and to provide overall control of the cbs system.

10.9.2 Trunking Systems

A logical extension of the common base station, a trunked system provides users sharing a base station to call on a pool of radio channels rather than each user having a dedicated assignment. A channel is assigned to the user for the duration of the call, whereupon it is then automatically available to another user. When all channels are in use, further calls have to wait until a channel becomes free. In a simple 'pure loss' system, calls attempted when all channels are occupied are rejected and the user must attempt the call again until a channel is available. An improvement in the grade of service and spectrum utilisation can be achieved by the use of queueing, where requests for channels are serviced under microprocessor control. Rotational or 'first-in-first-out' (FIFO) queueing, deals with calls in the order in which they are attempted whilst with random queueing, calls are put in a memory and given access to a free channel in a random order.

A typical system would involve between three and ten radio channels with around four users per channel each having between 20 and 100 mobiles. The potential improvement in spectrum utilisation will depend on the number and size of the users' fleets. Typically a gain of around 2.5 could be expected relative to a shared single channel system, although advantage could be taken of further improvements by creating groups of users having unbalanced traffic.

Supervision of the trunked system is maintained through the use of control signals which may be transmitted on one of the information channels or, where six or more channels are involved, by the use of a dedicated data channel.

Users are being encouraged to introduce trunking as a means of making better use of existing channels, whilst emphasis is being placed on the use of these techniques in new frequency bands becoming available to mobile radio.

10.9.3 Data Systems

The increasing use of microprocessor based systems and greater awareness in the field of information technology has lead to an increased demand for the use of land mobile data systems. Short coded messages may be used on conventional speech channels for selective calling, vehicle identification, routine information and status reporting, and offering considerable savings in 'air time' over conventional speech systems.

Continuous data streams for transmission of long coded messages, computer data or, in the future, digitised speech signals will require dedicated radio channels.

In a move towards the standardisation of digital transmission systems, the Radio Regulatory Division has issued a Code of Practice (4) for a recommended general purpose digital signalling system for short coded messages using Fast Frequency Shift Keying (FFSK) as the data modulation method and a data rate of 1200 bits/sec. The system will find a ready application for the supervision of future trunked mobile radio systems.

10.10 FREQUENCY ASSIGNMENT TECHNIQUES

Within the p.m.r. bands between 70 MHz and 470 MHz, around 900 frequency channels are currently available in the UK although not all of these may necessarily be used in a given location. It is important therefore that a frequency be re-used as many times as possible throughout the UK without mutual interference being experienced from one system to another.

Maximisation of the frequency re-use requires the use of detailed topographical data, a representative propagation model, knowledge of the field strengths necessary for communication and the degree of protection necessary against interfering signals, to achieve the required grade of service.

The choice of frequency channel necessitates, in addition, the storage of technical and administrative data on all existing p.m.r. users.

The large amount of data required, the complexity of the calculation processes involved in making a frequency assignment and the need to ensure efficient use of the available frequency spectrum has dictated the use of computer techniques.

Following the development of the first such assignment system by the Directorate of Radio Technology in 1974, a more efficient and equitable use of the land mobile frequency spectrum has been achieved in the UK. In order to meet the increasing demands of the p.m.r. service, to reduce the amount of manual intervention in the frequency assignment process, and to reduce the substantial computer processing time involved, a new automated frequency assignment and licensing system is to be introduced in 1985. Skiffins (5).

The system is based on a Perkin Elmer 3252 processor, configured with 2 Megabytes of MOS memory, a powerful floating point processor, and a 2k word writable control store used to provide a microcoded Fortran library. More than 1000 Megabytes of data storage is provided across five disc drives. The system provides for representative field strength contours to be generated and contour maps stored for each existing p.m.r. base station in the UK. During the assignment process the computer overlays the contour map of the proposed base station on those already stored for existing base stations in the same geographical area in order that an assessment of the mutual interference can be made. The process is complicated since some areas may be overlapped by more than one existing station. The process is repeated on each of the available channels until one is found meeting the appropriate assignment criteria.

Where users are required to share a frequency channel in the same geographical area, an assessment of the overall channel loading at the busiest times of the day is performed by the computer in order that a suitable channel can be found.

The operational requirements of more than 120 business categories have been identified, to ensure that as far as is practicable, an acceptable degree of protection is afforded.

Taken together with comprehensive licensing, accounting and management information systems, the computer will make an important contribution towards the efficient and equitable use of the private mobile radio spectrum in the UK.

10.11 FUTURE TRENDS

As telecommunications networks become more advanced, the use of new developments in technology will have a significant impact on the requirements and possibilities for mobile radio. The size, weight and power consumption requirements of portable radios will be reduced with the advance of VLSI, and development of surface acoustic wave (SAW) technology.

Portable radiotelephones operating within a small cell structure will become commonplace and instant communication with people on the move will be taken for granted.

Greater integration of line and radiocommunications networks will be required in order to allow development of advanced mobile communications.

So where will the radio spectrum come from to meet this growing demand?

Advances in technology which allow the exploitation of new and higher frequency bands may provide part of the answer particularly in the development of satellite communications networks for land mobile services. A move away from terrestrial broadcasting as greater use is made of the new infrastructure offered by cable technology and the advent of direct satellite broadcasting (dbs) systems may in the long term provide relief in the bands below 1000 MHz. In the medium term however, attention must be paid to advancing narrowband technology as a means of further exploiting limited spectrum resources. In particular single-sideband modulation and speech encoding techniques could offer substantial savings for the mobile radio service.

10.11.1 Single-Sideband Modulation

In order to evaluate the potential for 5 kHz single sideband, the Directorate of Radio Technology has carried out a series of subjective assessments of different forms of modulation with the aid of industry, universities and users of private mobile radio. Durkin & Barnes (6). The tests were designed:

(i) to establish the relative subjective quality of ssb, am and fm systems under field trial conditions by regular users of private mobile radio.

(ii) to compare the frequency re-use parameters of ssb, am and fm systems under field trial conditions,

(iii) to establish the relative interference potential and immunity of transmitters and receivers of ssb, am and fm systems,

(iv) to establish the long term frequency stability of ssb equipment with particular reference to hand portables and to compare the effects of frequency drift found in practice on the three systems.

(v) to compare the performance of the four systems for use in wide area coverage, e.g. quasi synchronous operation.

The tests showed that normal users judged the 5 kHz pilot-carrier ssb system to give a standard of service at vhf of a similar order to that of the present 12.5 kHz am and fm systems in use in the UK. The co-channel performance was such that the re-use factor of ssb was not inferior to that of the present 12.5 kHz systems, and the adjacent channel performance only slightly inferior to existing am equipment. Overall it was concluded that if 5 kHz ssb equipment could be made available at a comparable cost to that of existing equipment, the channel capacity per MHz could be at least doubled.

Recent developments in feed-forward automatic gain control systems and phased-lock-loop technology could lead to an economically viable ssb system suitable for the land mobile radio environment.

10.11.2 Digital Encoding of Speech Signals

Advantages that can be gained by representing speech waveforms in digital form include:

- Resistance to interference and distortion by the use of regenerative repeaters.
- Ease of handling in modern electronic switching systems.
- Speech, data, and telegraphy can be fully integrated in the transmission network.
- Greater privacy due to easy encryption.
- Compatibility with data processing systems.

Many methods of converting analogue speech signals to digital form exist. Waveform coding techniques such as p.c.m. or delta modulation offer high quality speech resolution but require a high data rate in the order of 30 to 60 kbits/sec. For narrowband applications, speech synthesis systems offer the greatest advantages, giving an acceptable speech quality at very low data rates (1 to 2 kbits/sec). In particular, linear predictive coding (lpc) is emerging as the most likely possibility for mobile radio systems, following the rapid development of integrated lpc speech synthesis circuits. Linear predictive coding approximates the vocal tract by means of a multistage recursive digital filter which produces an output approximating the excitation function needed to construct the speech waveform. The digital lpc signal is sampled and an estimate of the present sample is computed from previous samples. An error signal derived from a comparison of the present sample with the estimated sample is used in the waveform re-construction process.

The increasing need for rapid and sophisticated mobile communications together with the requirement for integrated communications networks will present a challenge to the design of future land mobile radio systems. The rapid developments in communications technology will ensure that such future systems are realisable. Careful planning will be required both nationally and internationally to ensure that future developments may be accommodated within the limited resources of the radio-frequency spectrum.

REFERENCES

1. Independent Review of the Radio Spectrum (30-960 MHz) - Interim Report 1982. HMSO Cmnd 8666.

2. Terrestrial Land Mobile Services: Spectrum requirements and availability to the end of the century. Report of the Mobile Radio Committee HMSO 1982.

3. Report on the spectrum requirements of Private Mobile Radio Services in the UK, 1975-1999 - Home Office 1975.

4. Code of Practice for the transmission of digital information over Land Mobile Radio Systems - MPT 1317 -Home Office 1981.

5. Skiffins RM, 1983 - Frequency Management in the United Kingdom for Private Mobile Radio Systems -Convegno 'Servizi Radiomobile Terrestri' - Bordoni Foundation, Bologna, June 1983.

6. Durkin J and Barnes D M, 1982, An Evaluation of Single Sideband Radio for use in vhf Private Land Mobile Radio Services - Proc 32nd IEEE Veh. Tech. Conf. San Diego.

7. Bands I and III - A Consultative Document 1984. -HMSO Cmnd 9241.

8. Final Report of the Independent Review of the Radio Spectrum (30 - 960 MHz), 1983. - HMSO Cmnd 9000.

9. MPT 1323 - Performance Specification for Angle Modulated Radio Equipment for use at base and mobile stations in the pmr service operating in the frequency band 174 - 225 MHz. - DTI 1983.

10. Proposals for the Exemption from Licensing of Four Categories of Low Power Radio Devices - A Consultative Document. DTI 1984.

TABLE 10.1. Summary of predictions

Category	1980		1985		1990		2000	
	1975 Projection	Actual	1975 Projection	New Proj	1975 Proj	New Proj	1975 Proj	New Proj
Central Govt	3,500	5,700	5,250	7,500	7,000	9,500	14,000	13,500
Local Govt	31,600	29,900	40,000	35,000	46,500	40,000	61,000	50,000
Manufacturing Commercial	96,100	99,540	130,250	118,650	165,000	136,000	225,000	147,000
Services	12,500	39,231	15,500	56,190	20,500	70,900	27,100	84,750
Public Transport	21,850	19,200	39,775	39,575	53,000	49,500	59,000	58,500
Other Transport	29,000	55,414	34,000	96,519	38,000	181,619	44,000	181,619
Utilities (excluding Post Office)	59,500	59,249	72,000	73,866	86,500	74,450	123,000	91,905
TOTAL	254,050	308,234	336,775	427,300	416,500	561,969	553,100	627,274

TABLE 10.2. Channel loadings

Category	Channel loading (Erlangs)	W/H	Ave mobiles per channel
Local Govt	0.57	1.35	94
Manufacturing/commercial	0.59	1.42	48
Services	0.55	1.24	67
Public Transport	0.50	1.00	122
Other Transport	0.50	1.00	44
Public Utilities	0.50	1.00	66
Other	0.54	1.17	74

MHz
174 — 235

<table>
<tr><th colspan="3">Allocation to Services</th></tr>
<tr><th>Region 1</th><th>Region 2</th><th>Region 3</th></tr>
<tr>
<td rowspan="3">174 — 223
BROADCASTING

621 623 628 629</td>
<td>174 — 216
BROADCASTING
Fixed
Mobile

620</td>
<td rowspan="3">174 — 223
FIXED
MOBILE
BROADCASTING

619 624 625 626 630</td>
</tr>
<tr>
<td>216 — 220
FIXED
MARITIME MOBILE
Radiolocation 627</td>
</tr>
<tr>
<td rowspan="2">220 — 225
AMATEUR
FIXED
MOBILE
Radiolocation 627</td>
</tr>
<tr>
<td rowspan="2">223 — 230
BROADCASTING
Fixed
Mobile

622 628 629 631
632 633 634 635</td>
<td rowspan="2">223 — 230
FIXED
MOBILE
BROADCASTING
AERONAUTICAL RADIONAVIGATION
Radiolocation

636 637</td>
</tr>
<tr>
<td rowspan="2">225 — 235
FIXED
MOBILE</td>
</tr>
<tr>
<td>230 — 235
FIXED
MOBILE

629 632 633
634 635 638 639</td>
<td>230 — 235
FIXED
MOBILE
AERONAUTICAL RADIONAVIGATION

637</td>
</tr>
</table>

Fig. 10.1 Excerpt from table of frequency allocations

Chapter 11

Switching and signalling

A.R. Potter

11.1 INTRODUCTION

There are now over 29 million telephones in the UK, with, in addition, the public switched telephone network (PSTN) providing automatic access to nearly all of the 500 million odd telephones around the world. All public radiophone systems require interconnection with this network if they are to provide a comprehensive voice communication service to their customers. This chapter therefore describes the structure of the telephone network and highlights its capabilities and limitations with regard to radiophone interconnection.

11.2 STRUCTURE OF THE TELEPHONE NETWORK

The principal factors which, historically, have dictated the dimensioning of telephone networks have been the relative cost of switching and transmission equipment. Each telephone is connected to its local exchange by an individual pair of copper (or occasionally aluminium) wires with analogue transmission of the speech occuring in both directions simultaneously. Cables are relatively well balanced with respect to ground to ensure crosstalk coupling is kept to a very low level (specification is better than 100dB of crosstalk attenuation). The geographical area served by the exchange is dictated by cost of the cabling and transmission loss considerations; a typical local exchange generally serving a radius of approximately 5km, although this figure is less in urban environments and greater in some rural areas where heavier gauge cables or amplified circuits may be used to serve remote customers. A mixture of cable types is used between the exchange and the customers; nearest the exchange the finest gauges (typically 0.3mm) are employed with several hundred pairs in a single cable sheath, while closer to the customer the heavier gauges are used. Overall it is necessary to keep within a maximum transmission loss of 10dB(at 1600Hz), while the DC loop resistance of the line is also a limiting factor for signalling between the telephone and the exchange. Older exchanges are limited to lines of 1000 ohm loop (plus telephone) while the modern exchanges can work with lines of up to 1500 or 2000 ohms when this factor no longer becomes a significant constraint.

There are some 6300 local exchanges in the UK and economic considerations make it not viable to fully interconnect these. Thus an hierarchy of trunk exchanges is used to provide the correct balance between transmission and switching. Direct routes are used between local exchanges where the distance and quantity of telephone traffic justifies this. Otherwise intermediate switching centres permit traffic from other sources to be assembled together onto a common group of circuits to a distant exchange.

Intimately linked with this hierarchical structure is the overall transmission loss strategy for the network. With the existing telephone apparatus, the maximum loss that can be tolerated is some 33dB and this has to be allocated between the switching and transmission systems at various levels in the network. In practice the maximum loss would be unacceptable if it were encountered by the majority of telephone connections and network design ensures that the majority of connections suffer a transmission loss well below this limiting figure.

Fig.11.1 shows the existing structure of the UK telephone network. Most of the long distance traffic is carried by the trunk network using two-wire Group Switching Centres (GSCs) of which there are approximately 350. Since

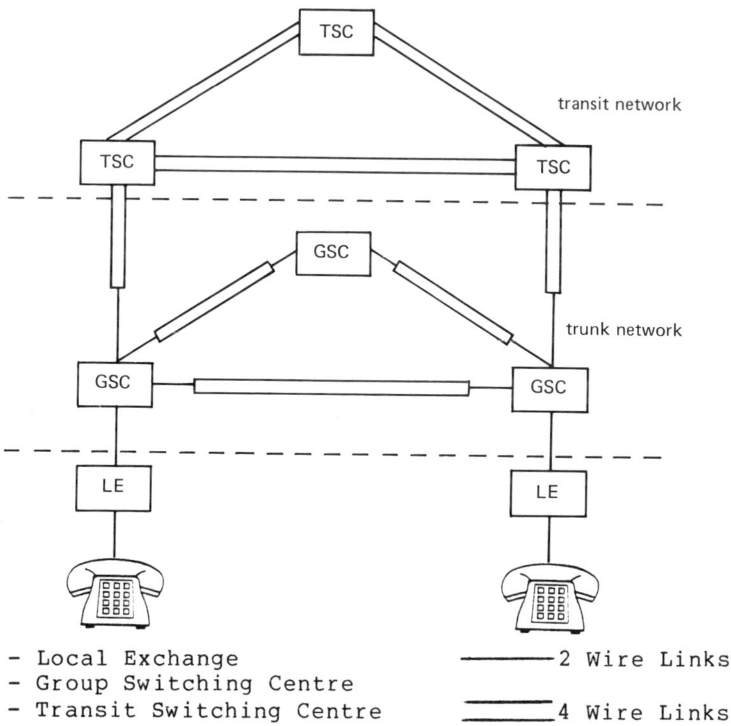

Fig.11.1 Structure of the UK Telephone Network

2-wire switching is employed with, in general, 4-wire transmission between the GSCs, the minimum loss practicable for the transmission links between the GSCs is 3dB - this is necessary to ensure adequate stability margin around the 4-wire loop of such a transmission system. This, together with the 'switching' loss encountered in the GSCs themselves, limits the network to the use of a single GSC for tandem switching of trunk calls. For this reason GSCs only provide for sufficient routing digit translation from the trunk code (the ABC digits of the national number (0) ABC XXXXXX) to route through one intermediate trunk exchange. The restrictions of this approach in terms of providing full automatic trunk access across the country have resulted in the provision of a so-called Trunk Transit Network of 4-wire analogue exchanges interconnected with 4-wire transmission links and high speed signalling systems. Up to 3 intermediate transit switching centres may be linked in tandem to provide a connection whose loss is constant.

11.3 TRANSMISSION AND SWITCHING

Historically transmission and switching were separate disciplines of telecommunications with perhaps the only common feature between them being the need to provide a 3.4kHz voice circuit interconnection. The advent of digital techniques has broken down many of these barriers and will have a far-reaching impact on the telecommunications network.

As indicated earlier, 4-wire techniques, where each direction of transmission travels separately, is used extensively for trunk transmission. This permits amplification of the signal to compensate for the build up of cable loss and thus, to a first approximation, a loss which is independent of distance can be achieved.

To permit efficient utilization of expensive long distance circuits, multiplexing of many speech channels onto a single cable is used. Using amplitude modulation of the baseband analogue signal with the carrier suppressed and only a single sideband, channels are stacked at 4 kHz intervals in a frequency division multiplexed (FDM) grouping. FDM systems in the UK today typically carry 960 or 2700 channels on coaxial cable systems operating at up to 4092 kHz or 12435 kHz respectively. A 61160 kHz system also exists providing 10800 channels.

Digital coding of speech using an 8 bit representation with sampling of the analogue speech signal at an 8kHz rate has been adopted for many years now as an international standard for telephony transmission. The repetitive nature of the sampling process permits the interleaving of samples from other speech channels - a process known as time division multiplexing (TDM) - which is illustrated in Fig.11.2 . The process of digital coding used is known as pulse code modulation (PCM). A TDM grouping of 30 speech channels with two additional time slots being allocated for signalling and timing information gives a digit rate of 2048 kbit/s. This is the basic primary multiplex building block for digital transmission systems.

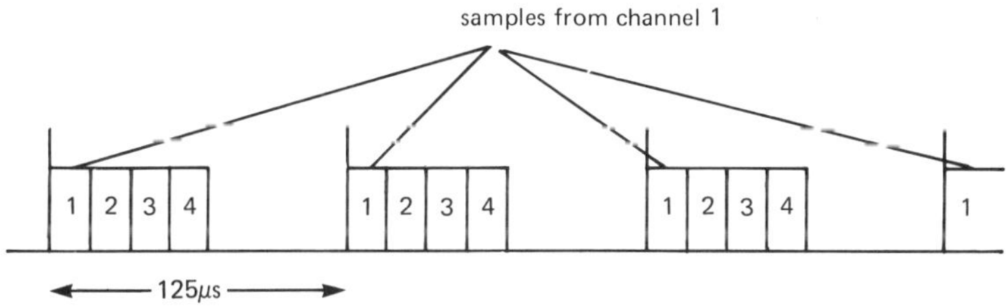

Fig.11.2 Principle of Time Division Multiplexing

Significant application of 2048 kbit/s digital transmission systems has already been made in the UK. The principal useage has been in the so-called junction network for transmission between local exchanges and GSCs. This uses copper pair cable with digital regenerators spaced at approximately 2km intervals. Obviously a major increase in the capacity of such junction cable systems, where previously baseband analogue transmission has been employed, is possible.

Higher order multiplexing of PCM 2048 kbit/s streams using TDM principles is then used for transmission on coaxial or optical fibre cable systems. The principal system in use provides 1920 circuits at a line rate of 139264 kbit/s although higher speed systems are under consideration.

Four different types of switching systems are in use in telephone exchanges in the UK today, ranging in technology from the Strowger system using 2-motion selectors invented nearly a century ago, to digital switching systems with computer control.

The basic Strowger system, which works by direct control of the two motion selector from the make-break pulses from the telephone dial, has been considerably enhanced over the years. Of these enhancements, the most significant has been the provision of registers which receive the dialled information from the customer and then translate it into the appropriate routing codes for access through subsequent exchanges. This has permitted common numbering schemes for large cities and standardised trunk codes to be used. In small exchanges, where the cost of such registers was prohibitive, the inflexibility of the Strowger system is apparent where routing codes and numbering schemes differ from exchange to exchange. Figure 11.3 shows a schematic of a typical small Strowger exchange and it should be noted that the directory number of a subscriber is directly related to the appearance of his line on the final selector.

A more up to date electromechanical system using the crossbar switch has been used in the UK from the mid 1960's. This system uses a matrix of relay type contacts which are operated by selecting appropriate row and column bars using electromagnets. Physically the crossbar switch assemblies can be quite large. However subject to certain routing

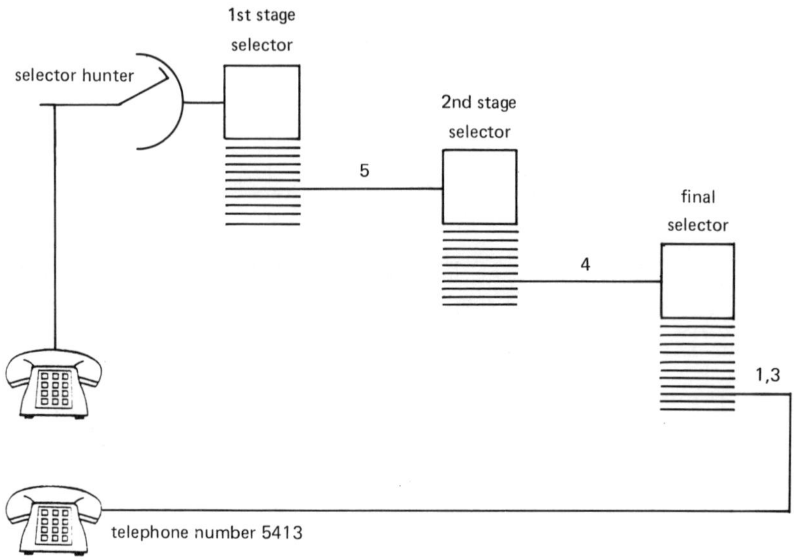

In the first and second stages of switching the selectors hunt across their outlets to find a free second/final selector.

Fig.11.3 Schematic of a small Strowger Exchange

constraints within the exchange, each column of the switch may carry a different telephone call. In addition, the relay contact switches are inherently more reliable than the Strowger two motion selectors which require routine maintenance.

The alternative modern electromechanical system to crossbar is the sealed reed relay for telephone switching. The reed relay switch has, typically, 4 reed inserts, surrounded by a coil which may be energised by a relatively low level current compared with the crossbar switch. This makes reed relay systems much more compatible with electronic control. The sealed reed relay also offers, in principle, the promise of the ultimate in reliability for an electromechanical switch arrangement. Unfortunately in practice this promise has not always been fulfilled because of the low operating and release forces involved. The first production reed relay exchange was introduced in the UK in the mid 1960's. This was the TXE2 system of which over 1300 units are now in service. Figure 11.4 shows a schematic of the TXE2 system and it may be seen that a centralised or common control arrangement is used. This works at two levels, whereby registers receive the routing information from the customers while the common control translates this and sets up appropriate paths through the switching matrix. A duplicated common control is employed, one in operation and the other acting as a standby.

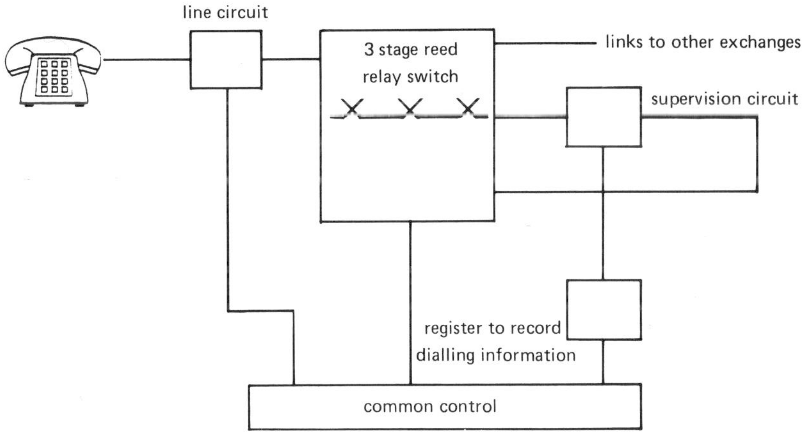

Fig.11.4 TXE2 Common Control Exchange

In the mid 1960's it was recognised that switching of the time division multiplexed digital highway produced by a PCM transmission system could offer significant economic and transmission advantages. Such a switching system is illustrated in schematic form in Figure 11.5 and relies on the use of relatively high speed digital read/write memories and switching circuits. The read/write memory provides for translation of the 8 bit PCM word from an incoming timeslot to a different outgoing timeslot and thus both space (as in traditional analogue exchanges) and time switching is employed. Each switching element also handles many speech channels multiplexed onto a single highway and thus compact and low cost exchanges may be realised with this technique.

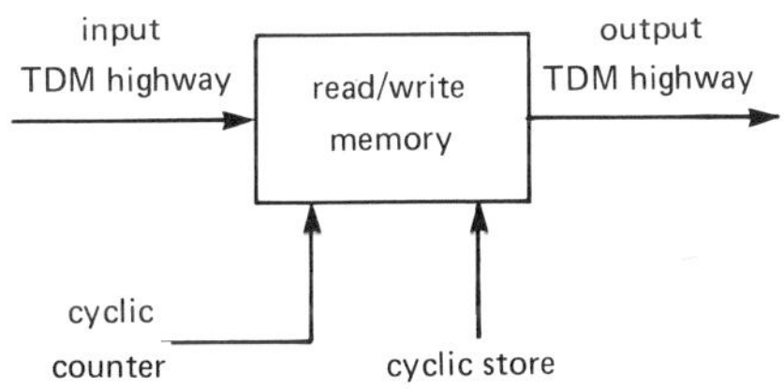

Write address controlled by cyclic counter stores pcm sample in fixed location.

Read address controls output timeslot into which sample is placed. Contents of cyclic store may be updated to vary the input timeslot.

Fig.11.5 Digital Switching - Principle of Operation of Time Switching

The switching function generally involves the two processes of traffic concentration and interconnection switching. Concentration is possible since in general the individual telephone lines entering an exchange (private exchange lines excepted) have relatively low occupancies. The occupancy or traffic is measured in Erlangs - named after the originator of the concept - which reflect the average occupancy of a circuit or group of circuits during a period (of usually one hour) measured during the busiest time of the day. Typically the traffic on individual subscriber lines is between 0.02 and 0.1 Erlang (taking both originated and terminating calls) and concentration ratios of up to 10:1 may be readily achieved in the first stages of the switch. In every case certain traffic measurements have to be made to ensure that only a very small proportion of calls encounter all links being busy and the overall target for a switching system is generally better than 0.02 blocking probability during the busy hour.

Interconnection switching, or route switching as it is termed, is then carried out and involves the establishment of links between large numbers of relatively high occupancy channels. In a digital switching system many hundreds of channels may be time division multiplexed into a single highway and as a result switch assemblies may be constructed which have very low probability of blocking even at relatively high channel occupancies.

Modern digital switching systems are invariably computer controlled and the UK's System X is no exception to this. Both local and trunk exchanges use a highly reliable multiprocessor arrangement with separate common channel, high speed, signalling links between the exchange control units. The use of software driven stored program control permits the very complex algorithms necessary to allow a wide range of advanced facilities to customers to be implemented. In addition an inherent flexibility exists to add new services and facilities in the future by updating the system software.

The ability to switch the digital transmission system without demultiplexing and demodulating back to analogue baseband has significant transmission advantages and ultimately will permit a network whereby a fixed loss from local exchange to local exchange will result. Already the UK is heavily committed to a program of installation of digital switching and transmission systems interconnecting all major centres on the integrated digital network. Within this integrated digital network the exchanges will run in a mutually synchronized manner. This is necessary to ensure that periodic loss of information on the interconnecting transmission links does not occur - not of itself too serious for speech but of significance if data is being carried. Synchronisation is achieved by exchanging information between exchanges, about the timing of the transmission systems entering an exchange relative to its own local timing source, and adjusting the exchange timing sources appropriately.

The advent of low cost integrated circuit PCM coder-decoders has also permitted the extension of digital switching to the subscriber line concentrating stages of local exchanges. Thus the latest generation of local exchanges are now fully digital in operation with analogue/digital conversion being carried out where the subscribers line enters the exchange. A feature of this approach is that the per line interface thus resulting may be varied to suit different types of terminals at the customers end.

A further development from the all-digital local exchange is the extension of the digital stream out to the customers premises using the existing pair of copper wires. Already the feasibility of transmitting the 64 kbit/s PCM channel, plus some additional capacity for timing and signalling, over several kilometers of local cable has been demonstrated. One technique of achieving this is a time divided arrangement whereby a word is transmitted at a relatively high speed - typically 512 kbit/s. When this is received by the customers' equipment the return word is transmitted, the whole process including round trip cable delay taking less than the 125 micro second repetition period.

With the integrated digital network the capability will thus exist for digital transmission from end to end across the network with a capability for either speech or high speed data. This new capability is being called the Integrated Services Digital Network or ISDN. Specialist ISDN equipment at the customers end will permit standardised data interfaces to be provided to terminal equipment. The likely standard for local transmission will be for n x 64 +16 kbit/s where n will normally be 1 or 2 and the 16 kbit/s residual channel may also be employed for packet type data as well as signalling, although in this case specialised access arrangements at the local exchange will be required. In addition higher speed (eg 2048 kbit/s) multiplexed assemblies will be transmitted on higher quality cable or radio systems in the local network to permit digital private exchanges access to the ISDN.

11.4 SIGNALLING

Traditionally, signalling within the network has been directly influenced by the ten impulses/second make/break output from the telephone dial. This directly controlled the Strowger two-motion selectors and was then repeated by signalling relay sets over audio junctions between exchanges. The advent of frequency division multiplexed long distance transmission systems brought in techniques of signalling between exchanges using in-band tones. In the UK the standard adopted is a single frequency 2280 Hz signalling arrangement although this still operates at 10 pulses/second.

The register controlled exchanges within the Transit network use high speed dual tone multi-frequency signalling whereby one out of six tones from each of a lower and upper band can convey a much wider range of information than the more basic single frequency systems.

In point to point PCM systems, timeslot 16 is employed to convey signalling information about each individual channel with 2 kbit/s of capacity being available for each.

Simple repetitive codes are employed rather than a message based system.

In the integrated digital network environment, a common channel signalling arrangement is employed with timeslot 16 providing a direct message link between controlling processors. Access is provided from these centralised units by paths which are semi-permanently switched through the exchange. The CCITT No 7 signalling system is the standard adopted. This has a basic 32 bit word structure with a signalling word of variable length of up to 69 bytes (1 byte = 8 bits). A signalling information field of between 3 and 62 bytes is used together with 5 bytes for various 'housekeeping' tasks. Two bytes are then employed as a cyclic redundancy check pattern and in the event of an error being detected the message is retransmitted. A single 64 kbit/s common channel signalling highway has adequate capacity for the information relating to many hundreds of channels. In addition, a much greater range of information can be transmitted across the network permitting such features as calling line identity and alternative routing to be possible.

At the subscribers end a dualtone multi-frequency system using 1 out of 4 frequencies at low and high bands has been standard for many years. This has seen widespread application in some countries but only recently has appropriate equipment been available to support its use in the UK in public exchanges. This system permits access to various advanced facilities at the exchange such as call diversion, short code calling to frequently used numbers, automatic alarm calls since additional codes outside the 10 basic routing digits are possible. To this end telephones with multi-frequency signalling are generally equipped with an additional two buttons labelled with * and #. These are used to appropriately punctuate information to the exchange, the * generally representing the beginning or division between various parts of a signalling sequence with the # representing the end. The * and # are also used to indicate 'activate' and 'cancel' respectively.

In an ISDN environment, signalling between the equipment at the customers end (the so-called Network Terminating Equipment or NTE) and the digital local exchange uses a new Digital Access Signalling System or DASS. DASS is based on High Level Data Link (HDLC) protocols and a compelled method of operation is employed where each signal transmitted is acknowledged as correctly received before the next interchange takes place. In addition to the normal routing information, DASS includes a service indicator code which says whether a telephony or data call is being made, the data service (Teletex, Videotex, Facsimile etc) and the data rate of the data terminal equipment attached to the NTE.

Digital private exchanges with an ISDN service capability may also be accommodated within the DASS protocols including calling line identity as well as the ability to work multi-link access procedures for the multiplexed channels available in such an environment.

11.5 RADIOPHONE SWITCHING AND INTERWORKING WITH THE TELEPHONE NETWORK

A mobile radio system with its associated switch may be considered as a communications network in its own right and its organisation is described in Chapter 12 on trunking and interworking. It is instructive to compare the organisation of such a system with its fixed counterpart since in a mobile system the switching is carried out at two different levels. The first of these is in the allocation of radio channels to the mobiles themselves. In this operation a relatively large (and somewhat unpredictable) number of mobiles have access to a significantly smaller number of radio channels. In the current systems these radio channels are frequency divided, although there is interest in the future in the possibility of using time division techniques with digital systems. This channel allocation procedure is analogous with the traffic concentration function of a conventional telephone exchange.

The second level of switching is then provided by the so-called Mobile Switching Centre or MSC which provides for the interconnection between radio channels and links to the telephone network. In addition interconnection between mobiles via the MSC may be carried out and also direct linking between MSCs is possible. The overall radiophone network structure thus resulting is illustrated in Figure 11.6.

The switching and control aspects of an MSC are broadly similar to those of a conventional telephone exchange although the special application does produce some differences. As has already been pointed out the switching problem is reduced to that of interconnection rather than traffic concentration as it is traditionally carried out.

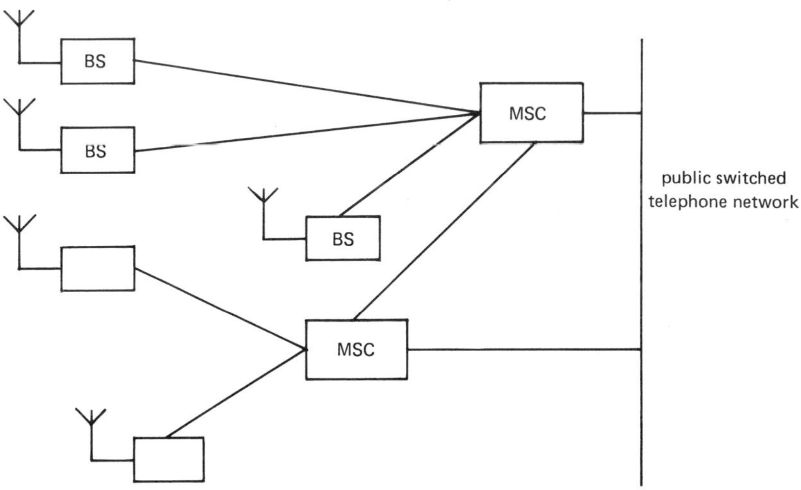

BS - Base Station   MSC - Mobile Switching Centre

Fig.11.6 Radiophone Network Structure

MSCs are invariably 4-wire switching systems and although those serving the earlier automatic radiophone systems use analogue switching, increasingly digital switching is being used. In these latter systems with their high availability time division multiplexed highways, traffic dimensioning within the switch is generally of little importance. Standard 8 bit PCM coding of the speech within the switch is adopted and digital connection between MSCs and the integrated digital public telephone network is a highly attractive proposition.

The principal differences that exist between the organisation of an MSC and a conventional exchange is in the operation of the control. For originated calls on the radiophone system, a mobile unit has to find and seize a free channel. This should be done with the MSC providing the overall control of this process to establish whether the mobile is allowed service, what facilities it may require and to ensure so far as practicable that the most appropriate channel is allocated.

Calling of the mobile requires the 'paging' signal to be sent from appropriate base stations to which the mobile will respond if able to receive calls at that time. If the location of a mobile relative to an individual base station is not approximately known then paging around a number of base stations is necessary with the attendant time and signalling overheads that result. Finally, while calls are in progress the cellular radiophone systems can 'hand-over' a call with a short break in transmission to an adjacent base station as the mobile crosses the boundary between the two service areas. This requires additional switching and signalling actions which must be carried out rapidly (generally in less than 1 second) if a significant interruption to the conversation is not to happen.

When interconnecting between a radiophone system and the telephone network it is particularly important to give proper consideration to the numbering and routing strategies together with the attendant signalling interfaces to be provided. In addition the transmission plan and attendant strategies require serious attention if significant degradation is not to be introduced above that already encountered in telephone connections.

If a radiophone system has a small number of users and only a single MSC (with the resulting long interconnecting private circuits to its base stations in distant areas of the country) the problem of numbering and routing to access that network is relatively simple. However, once the number of MSCs grows then careful planning and organisation is needed. Allocation of an alternative trunk code for each MSC is highly inefficient and can only be justified if significant spare numbering capacity exists. The most efficient is to allocate a single number group (which may be a trunk code or even a subset from that). This then implies that calls from the telephone network are routed to the nearest MSC and the network of MSCs then look after identification and location of the required customer.

The form of signalling used depends both on the level

within the telephone network hierarchy that the interface to the radiophone systems is provided and on the type of equipment available at each end. Interestingly one of the principal constraints is the requirement of call charges to be allocated between the two networks. Modern digital systems with common channel signalling minimise these problems but for some years parts of the telephone network will only be able to provide the older types of signalling interface.

The dynamic range of the levels of the speech signals on the current telephony network present major problems in interworking with mobile radio systems. In the limit the variation betwen minimum and maximum loss can be as much as 33dB and in addition, the dynamic range of the speech signal itself and the variation in individual talker levels must also be taken into account. Overall this has led to telephony networks needing to operate with a total dynamic range approaching 70dB. Even this range is designed to accommodate a certain clipping probability for the combination of loud talkers and low loss connections.

The radio channels used for radiophone cannot support such a wide dynamic range and typically the UK Cellular (TACS) system is designed to accommodate FM deviations of up to 9.5 KHz with a range of approximately 30dB. Companding (literally, compression/expansion) techniques are used to reduce this dynamic range variation at the entry and exit points within the radiophone network. Typically, however, companding ratios of only 2:1 can be achieved with adequate linearity at the present time and in practice therefore it is necessary to design the transmission levels at the interface between the telephony and radiophone networks to allow for a higher clipping probability. Without taking account of the degradations that the radio channel itself can introduce, such as fading and co-channel inteference, a greater proportion of calls will, however, suffer additional degradation than encountered on the telephony network. The progressive digitization of the network will do much to reduce these problems in the future - within 10 years more than 50% of trunk traffic will be handled on all digital connections. Even so it is apparent that it would be naive to expect the same level of quality from a radiophone connection compared with that provided by the telephony network.

BIBLIOGRAPHY

1. Telecommunications Networks edited by J E Flood (Peter Pereginus Ltd (IEE)) - General background in switching and transmission.

2. Post Office Electrical Engineers Journal (Now retitled British Telecommunications Engineering) published quarterly. Various issues over last 15 years concerning UK switching and transmission systems.

3. IEE Colloquium - Recent Developments In System X. London, April 1982. A number of papers describe the ISDN.

Chapter 12

Trunking and interworking

R.W. Gibson

12.1 INTRODUCTION

The aim of a radio trunking system is to allow many users to share a group of channels in a way which allows the best grade of service even when traffic is high.

The basic configuration of a trunking system is shown in Fig. 1. The base station is able to transmit and receive on any of the channels available to the system. It receives requests for service from users who may gain access by radio or by land line and allocates channels accordingly. If only radio access is required the system serves as a multichannel repeater. If land line connections are made to the fixed stations a switching matrix will be needed to connect any of the land lines to any of the radio channels.

The central processor has 3 functions: to receive requests for service; to process them, forming an ordered queue of waiting calls if necessary; and to allocate free channels by signalling to the mobiles concerned and making connections through the switching matrix.

When all channels are busy further requests for service can be dealt with in one of two ways:

(a) Pure loss Calls are merely turned away and the user tries again later.

(b) Queueing Calls are put into a queue to be dealt with in some chosen order.

A "first come first served" queue is the most efficient method. Pure loss and queueing with random order of service both increase the chance that users will suffer long delays and so waste channel capacity. The waste is from 5 to 15% for random service and larger for pure loss, depending on system size. An effective "first come first served" queueing system is therefore an aid to efficient channel use.

In practice the queueing system is usually modified from strict first-come first-served to allow various priority levels to be allocated. One is then deliberately increasing waiting times for some users or circumstances to allow others a better service.

The traffic on private mobile radio channels is significantly different to telephone traffic and these differences affect both the design and performance of radio trunking systems. In particular:

(a) The mobiles usually work simplex
(b) Conversation lengths are shorter (typically 15 sec mean call length compared with 3 mins for telephony)
(c) There are fewer users on the radio system.
(d) Often the users work in 'fleet mode' in which mobiles talk only, or mainly, to a dispatcher. In these circumstances the mobile may find the system 'busy' even when there are free channels because the dispatcher is talking to someone else.

References 1 and 2 give a detailed account of the effects of these considerations on the grade of service the user sees.

An ideal trunking system would only transmit useful information; channels would only be assigned for each 'word' spoken. Although such schemes are used on very high density telephony traffic they are not feasible for small radio systems. For duplex schemes a channel is normally assigned for the full duration of a conversation. For simplex schemes there is a choice between assigning a channel either for the full conversation or for just the burst of speech in one direction. The latter mode is theoretically more efficient but will tend to cause conversations to be 'broken up' by delays in obtaining a channel. Some manufacturers offer a choice of mode but for most applications assignment for the complete conversation is preferred.

12.2 GRADE OF SERVICE

In a 'pure loss' or 'blocking' system the grade of service is simply given by the probability that a call will be accepted. In a system having either automatic re-try or a queueing system it is possible to calculate or measure by simulation the statistics of the delays in obtaining service. (Note that it is usually impractical to measure the statistics on a real system under normal working conditions). Typical results of such measurements are given in Figure 2. This figure shows the dramatic improvement that a trunked scheme can give compared with a single channel approach. It also shows that different signalling schemes can have a considerable effect on the grade of service that a trunking scheme can provide.

12.3 CONTROL CHANNELS

Various techniques are used for the signalling involved in setting up calls. The three commonest methods are

(1) 'All-channel' signalling
(2) A 'Dedicated' Control channel
(3) 'Scanning' by the mobiles to find the current control channel

'All-channel' control schemes transmit the control signals either above or below the audio band and have been used with signalling rates up to 300 bits/sec. Such schemes can give a good basic system with adequate access times but are not as flexible as the other techniques. The other two schemes normally use signalling rates in the range 1200 to 4800 b/s and the messages are protected against errors by redundant coding and repeated as necessary to ensure reliable reception.

In a system having only a few channels the use of a dedicated channel for control purposes is clearly very wasteful and the scanning technique is appropriate.

In larger systems a dedicated control channel becomes desirable because of the importance of signalling time. Briefly stated the problem is that the time which elapses between a channel becoming free and its being allocated must be short compared with the rate at which channels become available. Otherwise a queue of unused channels builds up merely because the signalling is not fast enough. For example, in a 10 channel system with an allocation time of 2 secs. almost half of the system capacity is lost in this manner.

The time taken to assign a channel includes not only the time for the exchange of messages but also, in a scanning system, the time taken for the mobiles to locate the signalling channel. The latter factor becomes larger and larger as the number of channels is increased and eventually it becomes more efficient to fix the signalling channel more-or-less permanently. The point at which the loss of capacity due to signalling efficiency is greater than the loss due to under utilisation of the signalling channel is dependent on the exact signalling technique used but can be as high as 10 to 15 channels. Fig. 3 shows the results of a very comprehensive computer simulation of a system. The figure shows the traffic that can be carried when 30% of the users are forced to wait at least 20 sec for their call (average call length = 2¼ sec). Also shown is the traffic which could be carried by an ideal system (with infinitely fast channel assignment) for the same grade of service: the differences represent the signalling overheads of the real system.

It can be seen that under these conditions the optimum transition point from a scanning to a dedicated system is at 12 channels, but that the transition is gradual and so a change anywhere between 10 and 14 channels would be acceptable. It is also clear that the performance of both schemes declines for a large number of channels. This occurs because the signalling protocols are not fast enough to deal with very heavy traffic loads and so there are delays in re-assigning free channels to speech. Even the dedicated scheme shows a loss in the speech-carrying capacity per channel for more than 16 channels, though the decline in performance is gradual, and can be avoided by changing to very fast signalling.

12.4 CONTENTION CONTROL

A critical part of a trunking system is the technique used to signal requests for calls from the mobiles. If two mobiles ask for a call at the same time their requests may collide and thus fail. At low traffic rates almost any scheme is satisfactory, but only a very carefully designed protocol can cope efficiently with heavy traffic. In fact some protocols become unstable at high loads and cause the whole system to become blocked.

The key to efficient working is to allow the central computer to have control over when requests are made. Polling systems under which each mobile is called in turn are usually not appropriate: they lead to slow access times and low control channel efficiency. Instead a variant of the ALOHA principle may be used (Ref. 4,5). The central controller periodically sends out an 'invitation to make a request' signal. Any mobile wishing to make a call responds to the invitation after a variable delay. In simple schemes the delay is just a random period but this only gives a control channel efficiency of about 10%. By using more sophisticated algorithms for choosing the delay it is possible to get both efficient use of the control channel and short access times.

12.5 CONTROL MESSAGE FORMAT

The message format can be designed to meet special requirements in a particular application but this will rarely be justifiable. The EEA general purpose data format is suitable for most schemes.

A typical message will be as shown in Figure 4. The 'information' contains such things as:-

Address of calling party
Address of called party
Change to channel x command
Acknowledgements
Short data messages etc.

Some signalling on the speech channels is also desirable (e.g. end-of-conversation message) and this can either be done with the same format or by slower out-of-audio-band signalling.

12.6 RADIO DESIGN FOR TRUNKING

The radio designs (base stations, mobiles, or portables) can be virtually the same as normal PMR equipment. Multichannel equipment is obviously required and will usually employ synthesisers. At the base station the radios will be controlled by a computer which can vary from a large mini computer down to a small microcomputer depending upon the system size. A special interface card may be used to switch the radio functions (squelch, Tx key etc.) but in later equipment where micro-processors are used as part of the radio the inter-face may already be available.

In the mobiles and portables a microcomputer is used to provide the 'intelligence' and signalling functions. With modern VLSI techniques these extra units may consist of only one or two extra ICs.

The only requirement on the radio for trunking purposes is that it can switch channels fairly quickly: changeover within 10 to 20 ms is desirable for efficient working in heavily loaded systems.

12.7 INTERWORKING

Any trunking scheme already contains most of the elements necessary for interworking with fixed telephone networks (i.e. address-ing and call request messages, audio switching, etc). For private radio systems interconnection demands

1) A Vox switch to connect the duplex telephone line to the simplex radio
2) A code converter to convert the telephony dialling scheme (pulses or DTMF) to and from the radio message format
3) An audio interface to match the audio input and outputs of the radio to telephone standards.

12.8 FACILITIES

Of necessity the equipment used in a trunking scheme has the ability to exchange control messages. It is then a fairly trivial matter to add a whole range of extra features. These may include:-

Emergency Alarm
Call transfer
Conference calls
Priority service
Short messages (e.g. status)
Call back

12.9 IMPACT ON USERS

We may distinguish between small users, who will operate a few mobiles on a 'common carrier' system, and large users who will have their own specially customised system.

The small user will find that his mobiles and portables cost slightly more than standard units, but this may be offset by lower base station costs. He gains a much better service because he is no longer swamped by bigger users on a shared channel. He will lose all the bother of a shared channel and, except for an occasional wait for a call to be connected, he will appear to be working on an exclusive channel.

The large user gains because not only can he use his channels more efficiently but also the system will handle the channel and base-site selection which normally has to be done by the dispatcher.

Both types of user also gain the ability to use the 'extra' features (telephone interconnect, call switching). Also the automatic queueing of calls is a help to a busy user and an aid to safety in a moving vehicle.

12.10 REFERENCES

1. DAVIS, C.K. and MITCHELL, R.F.: "Traffic on a shared radio channel", IERE International Conference on Land Mobile Radio, Lancaster, Sept. 1979.
2. DAVIS, C.K. and MITCHELL, R.F.: "Studies of small trunking systems for mobile radio", IEE Conference "Communications 78', April 1978.
3. DAVIS, C.K. and MITCHELL, R.F.: "Traffic handling capacity of trunked land mobile radio systems', IEEE International Conference on Communications, Boston, June 1979.
4. KLEINROCK, L.: "Queueing systems", Vol. 2. John Wiley & Sons, 1976 and references therein.
5. ABRAMSON, N.: "The ALOHA system - another alternative for computer communications", Proc. AFIPS Conf., 1970 Fall Joint Computer Conf., 37, p. 281.

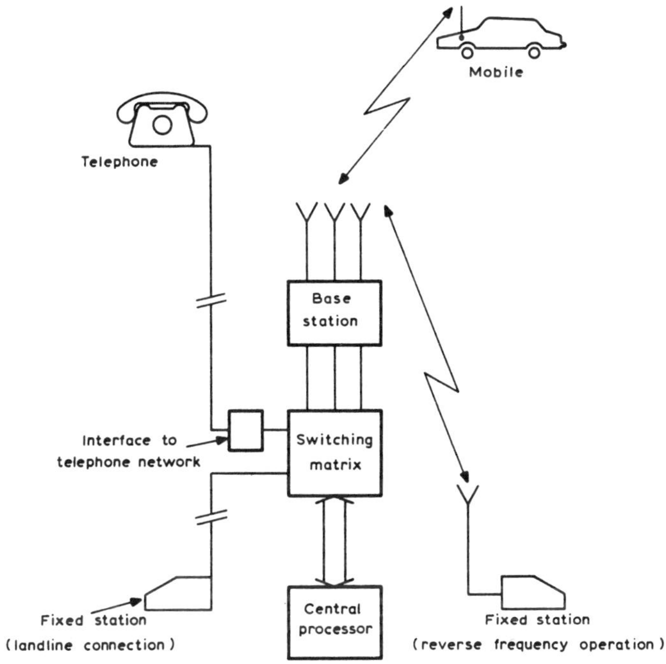

Fig 1 A basic radio trunking system

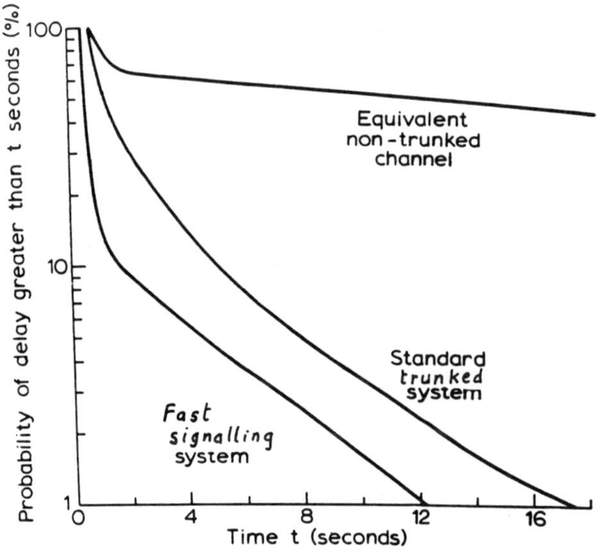

Fig 2 Channel Access Delay

(Channel loading 0.6 Erlang/Channel.
Mean call length 20 sec.
15 channel systems)

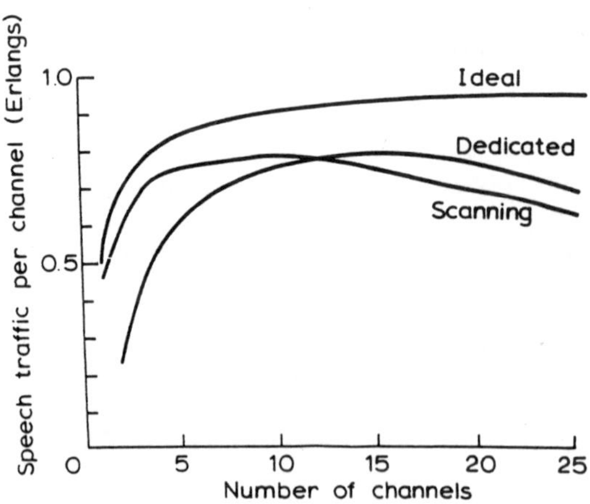

Fig 3 Traffic carrying capacity

(Mean call length 20 sec.
30% of calls wait 20 sec or more for access)

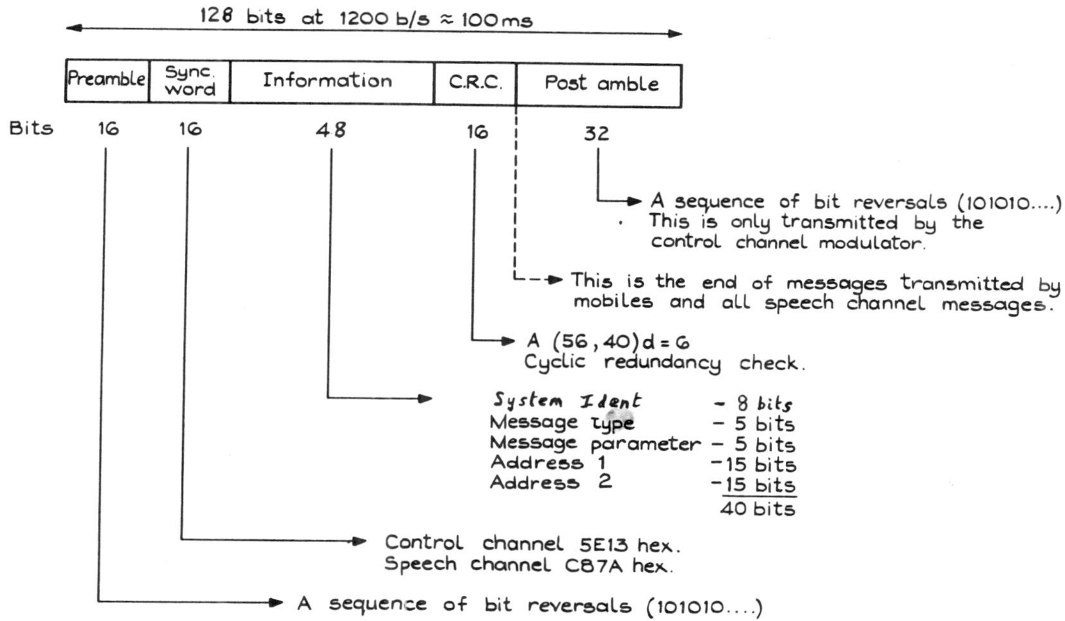

Fig 4 Typical Message Format

Chapter 13

Cellular mobile radio

J.M. Garrett

In this chapter we look at what makes cellular different from other radio systems and then consider some of the techniques that have been developed for cellular systems. The special features of systems in the field or under development are discussed in sections 7 and 8.

13.1 CAPACITY

There are radio telephone systems in many countries, providing service to a relatively small number of users and operating with about 100 channels from a number of base stations, generally chosen to give as wide a coverage area as possible. The capacity of such systems is limited by the number of channels available and the fact that they cannot be re-used except at base stations separated by very large distances. Growth in capacity requires extra channels - not readily available in today's mobile bands.

The main motivation for cellular was the need to provide high capacity mobile radio telephone systems that did not require very large numbers of channels. This leads to the adoption of base stations serving small areas or cells, so that the same channels may be re-used within relatively small distances. This is called **frequency re-use** and leads to two major considerations in the design of a cellular system, interference-limited operation and hand-off. We will examine both in more detail after a look at the organisation and structure of a typical cellular system.

13.2 SYSTEM ORGANISATION

As well as the mobiles, a cellular system comprises a number of base stations and a smaller number of mobile switching centres (MSCs) at which the radio channels may be connected to telephone circuits. The base stations, or cell sites as they are often called, are connected to an MSC by speech circuits and control or data links. Between the MSCs there are be speech and data circuits to allow calls to be forwarded across the cellular network.

Let us look at the events in a typical mobile originated telephone call on a TACS system. TACS, or Total Access Communications System, is the UK standard for use in the 900 MHz band. In the examples below it is assumed for clarity that there is only one TACS system in operation and that it uses combined paging and control channels.

When a mobile is turned on it scans the pre-determined control channels (whose identity is programmed into the mobile's memory) for the strongest control channel and then locks on to that to receive system status information. A customer making a call keys in the number required and then presses the mobile's SEND key, whereupon the mobile scans for a free access channel and attempts to seize it. Assuming successful seizure, the mobile is instructed to send its own identity and the required number and is then allocated a traffic channel, provided the mobile is known to the controlling MSC and allowed to originate calls. The controlling MSC routes the call into the Public Switched Telephone Network (PSTN) and records call details for call accounting purposes. When the mobile customer clears an ON HOOK signal is generated by the mobile and the traffic channel released. The mobile then resumes receiving on a control/paging channel.

A call to a mobile is routed by the PSTN to an MSC, which checks that the mobile number called is valid and that the mobile is not already engaged in a call. Because mobiles automatically register their presence as they move between paging areas, the mobile's location is known to within a paging area so the MSC instructs the base stations within that paging area to transmit a paging call. The mobile will receive the call from the base station to which it is tuned and respond on the reverse control/paging channel. This informs the MSC which base station to use at the start of the call. A free traffic channel on this base station is selected and the mobile instructed to switch to it. The mobile is then instructed to switch on the alert tone, (that the customer hears) and when the customer answers the call an OFF HOOK signal is sent by the mobile, the alert tone is switched off and the speech path connected. At the end of conversation the traffic channel is released and the mobile reverts to the control/paging channel.

In the above example it was assumed that the MSC knew the paging area in which the mobile was operating. This is because mobiles detect which area they are in by decoding the information sent on the control channels. When a change of area is detected the mobile seizes an access channel in the new area and sends a registration message to inform the system of its new location. In a system with several MSCs the registration message is forwarded to the mobile's home MSC so that it always has the last known location stored. Of course, if a mobile is switched off and crosses paging area boundaries then it cannot register and the location information will be incorrect for a time but calls to the mobile would fail anyway. However, as soon as the mobile is switched on again it will detect that it is has changed area and register so it is available once more to the system.

13.3 FREQUENCY RE-USE

Given a number of radio channels and an ideal plane earth to cover, there are a number of ways that the channels may be grouped into channel sets and allocated to patterns of cells that fit together and give continuous coverage. Some are illustrated in the figures.

For each pattern there is a distance D that separates the base stations using the same channel set. When divided by the cell radius R this is known as the re-use ratio.

It can be seen that the smaller the re-use ratio the fewer the channel groups that are required and, for a given total number of channels, the higher the trunking efficiency that may be achieved. The other side of the coin, however, is that the co-channel interference is increased as the re-use ratio is decreased.

13.4 CO-CHANNEL AND ADJACENT CHANNEL INTERFERENCE

The presence of interference is accepted in the design of cellular systems. The important criterion is the acceptable level of interference. This will depend upon:-

- the allowable carrier to interference ratio (C/I) for the static case
- a margin for fading due to terrain effects and multipath propagation

The C/I required for the static case is dependent upon the receiver bandwidth and modulation used. For 25 kHz spaced channels and 5 kHz peak deviation FM the typical receiver bandwidth would be 15 kHz and the relationship between C/I and signal-to-noise ratio of recovered audio may be approximated by the relationship:

$$S/N = 0.9\ C/I + 10 \qquad \text{for } C/I \text{ between 12 and 35 dB}$$

i.e. an "acceptable" signal-to-noise ratio of 21 dB may be achieved with a C/I of 12 dB, in the absence of fading.

To allow for the effects of Raleigh fading, as discussed in an earlier chapter, the equation above is modified to:

$$S/N = 1.0\ C/I - 9 \qquad \text{(for 99\% of the time)}$$

i.e. to achieve the 21 dB S/N requires a C/I of 30 dB, which is an increase of 18 dB on the static case. (Obviously a smaller increase would be allowable if the S/N requirement was reduced to a smaller proportion of the time.)

Companding, which is used in the audio path of several cellular systems offers a considerable improvement on the signal-to-noise ratio finally delivered and wider deviation systems can provide an acceptable signal-to-noise ratio for a lower C/I.

Fade margins of 10-15 dB are generally allowed, depending upon the certainty of coverage required and the type of terrain to be covered.

For systems, such as TACS and the US standard, that use wide deviation for speech and data there is also the possibility of interference from adjacent channels. This makes it desirable for an increased re-use distance ratio (resulting in a 12-cell repeat pattern) if there are prolonged periods of high deviation transmissions. However, for normal speech conditions it is acceptable to use a 7-cell repeat pattern.

Narrow deviation systems such as the German Netz-C, C-900 and MATS-E or the Nordic NMT do not give significant adjacent channel interference, which is thus not a factor in cell planning.

13.5 GRWOTH STRATEGIES

Whilst it might be possible to start a system with uniform cell sizes, there will come a time when extra capacity is required for part of the area covered. Cell splitting is used to provide the increased capacity. The new, smaller cells may be created by setting up new base stations (see Fig. 2) or by altering the coverage of the existing stations. The use of directional or sectorized antennas can give, for example, three cells where previously only one existed (see Fig.3). Such a technique saves the expense and difficulty of extra sites, although it does require directional antennae.

13.6 SWITCHING

Switching systems for cellular mobile services is frequently based upon established telephone switches that use stored program control. Indeed, some manufacturers claim the ability to run bith services together on one equipment. The functions required, line supervision, digit reception, analysis and translation, generation and reception of signalling, etc. are common to most types of exchange.

There are two particular areas of difference, traffic concentration and subscriber roaming. In a conventional switch for local exchange service there are a large number of subscriber lines connected but only a few active at any time. Much of the volume of a space-switched exchange is taken up with the per-line equipment and concentration stages of the switch. The cellular switch, however, is presented with concentrated traffic from the base stations. In effect the concentration stage takes place on air - i.e. there are a limited number of channels available in any one place and the switching system must handle these. It does not need to have equipment for each mobile subscriber.

On a conventional switch the subscriber terminals are connected to particular equipment positions, often identified by an equipment number (EN). The relationship between EN and directory number (DN) is often very flexible and may be under control of the subscriber, so that call diversion facilities etc can beinvoked. The same is true of switching systems for mobile service but there is also the added flexibility that terminals (mobiles) may move between areas controller by different switches and, indeed, be out of the service area altogether.

Cellular systems provide hand-off, as discussed in the following section, and this may be extended to effect hand-off between switches, where the configuration of cells served by different switches so requires.

13.7 HAND-OFF

A consequence of the use of small cells is that a significant proportion of mobiles will move across the cell boundaries during a call. Cellular systems are designed to provide automatic call hand-off from one base station or cell to another so as to maintain the transmission quality at an acceptable level. Typically, the steps involved in hand-off are:-

- base station monitors signal quality on a quasi-continous basis
- when quality is below a pre-set threshold the system instructs surrounding base stations to measure the signal strength and report
- system control selects new base station and channel at that station
- system instructs mobile to switch to new channel
- when mobile confirms its arrival on the new channel the system switches audio path to new base station and drops path to previous station.

These activities are all automatic and the telephone users will be almost unaware of hand-off, provided that the speech levels are maintained and that no appreciable break in speech is caused by the signalling and mobile's switching channel. The signalling techniques used by various systems are studied in the next section. Breaks in speech transmission may be as short as 250 ms or as long as several seconds, depending upon the signalling used. Subjective tests have shown that infrequent breaks of less than 250 ms cause little disturbance to conversation but frequent (say, every 5 s) breaks of this length or any breaks of >500 ms are readily detectable by telephone users and may be objectionable if other transmission impairments, such as noise and echo, are negligible.

Signal quality is generally measured by the use of a supervisory tone above the speech band. This tone may also have other uses (see section 7). In the case of the C-900 system the tone is replaced by a digitally modulated sub-carrier above the speech band, quality being assessed by measuring the jitter of the digital signal. Note that either technique allows the detection of poor signal-to-noise ratio, perhaps caused by interference, even though the signal strength may be high.

Signal quality is not the only possible criterion to determine whether hand-off is required. Distance or position measurement may be used alone or in conjunction with signal quality. The German C-900 system design makes use of the distance from base stations, as measured by the delay of digital signals continuously exchanged between mobile and base station. This elegant technique, which can locate a mobile to within about 0.5 km, offers the advantage of preventing the "dragging the cell boundary" by a mobile in a particularly good position that may occur if signal quality alone is used.

13.8 SIGNALLING TECHNIQUES

Some of the biggest differences between cellular systems may be found in the signalling techniques they adopt, despite the fact that, in principle, they all have to provide the same facilities! For example, they all need ways to:-

- give the system control over mobiles' frequency, power etc
- provide call control, with dialling, supervision etc
- obtain locations of roaming mobiles.

Some of these may be carried out automatically, whether or not the mobile is engaged in a call. Some occur only in the setting up or clearing down of a call. Others may best be carried out during a call, with as little interruption to speech as possible.

All have to deal with the hostile environment of VHF/UHF mobile radio channels in which fading, noise and interference may cause high error rates, so all have employed error protection and/or correction techniques to give satisfactory performance. We shall examine some specific techniques in detail.

For very low mobile speeds, such as a hand-held portable at slow walking pace, fading can cause the loss of many consecutive bits. At high vehicle speeds there may be insufficient correct bits between error bursts caused by fades to allow the errors to be "flushed out". Clearly, no amount of coding will allow satisfactory transmission when the mobile is stationary in a null!

13.8.1 NORDIC - NMT System

This uses audio frequency shift keyed 1200 bit/s binary signals with a convolutional code (by Hagelbarger) which allows correction of up to six consecutive bits in error provided there are guard periods of at least 19 correct bits between the errors. The basic data is divided into 64 bit blocks, to which are added 64 protection bits and a number of synchronisation bits as preamble, thus reducing the effective data rate to approximately 500 bit/s. NMT has chosen a compromise code that is suitable for use at 450 and 900 MHz over a pratical range of speeds.

13.8.2 TACS

TACS has followed the American standard in the use of a high basic data rate, in this case 8 kbit/s, which is Manchester encoded to 16 kbit/s direct frequency shift keying. Within this there are multiple transmissions, 5 for the forward control channel, allowing majority voting at the receivers. On top of this there is a forward error correction code that allows correction of up to two errors in each word of 28 wanted bits. The resultant data rate is still only approximately 1200 bit/s but with a very high likelihood of successful transmission. In fact the signalling will survive conditions that would be useless for speech.

13.8.3 C-900

The interesting feature of C-900 signalling is that there are two data channels provided, a fast and a slow, the slow one operating during conversation. The fast one, used for call control and distance measurement, operates at 5.26 kbit/s and uses forward error protection, reducing the effective throughput to approximately 2.4 kbit/s. The slow data channel is achieved by dividing the speech into segments of 12.5 ms each, which are time-compressed to about 11.4 ms, between which 4 bits of data are sent, thus giving a data rate of 320 bit/s. The significance of the slow data channel becomes apparent when one considers the activities involved in hand-off.

13.8.4 MATS-E

MATS-E also provides for a slow data channel by the use of an above speech-band sub-carrier of 4 kHz, modulated at 150 bit/s.

13.8.5 Signalling for Hand-off

Hand-off is one of the occasions when it is desirable to be able to exchange data between base station and mobile without interrupting speech. Another example would be for the sending of charge rate data during conversation or the invocation of "malicious call trace" without the caller's being aware of the request for trace.

NMT and TACS both suffer a brief interruption in speech transmission, with audio muting as soon as the mobile recognises that signalling is occuring. C-900, on the other hand needs no break in speech except during the switching to the new traffic channel and confirmation of such switching. Thus, potentially, such a system offers advantages for high density cellular systems.

13.9 TRANSMISSION QUALITY

In private mobile radio systems there are usually only a few connections to controlling operator positions and the speech levels on these circuits can readily be controlled to provide a satisfactory transmisson performance. However, when a radio telephone network in interconnected with a national telephone network, which is itself part of the international system, then there are several constraints on the transmission performance.

The most obvious factor is the high variability in speech level that may be encountered. Typical figures for mean speech voltage produced at the local exchange end of the line by a median talker on a 700-type telephone would be:

zero length	exchange	line	(SLR = 2 dB)	-15 dBV
median "	"	"	(SLR = 5 dB)	-18 dBV
limiting "	"	"	(SLR = 8 dB)	-21 dBV

In TACS the relationship between speech voltage at the base station transmitter input and receiver output and frequency deviation on the radio path is set at:

-25 dBV at transmitter input gives +/-2.3 kHz peak deviation

and

2.3 kHz peak deviation gives -15 dBV from receiver output.

The variation between quiet and loud talkers may amount to about 20 dB difference in mean level, and the variation in loss between different paths through the telephone network may also amount to some 20 dB. In practice, when speakers encounter a poor circuit they tend to speak louder than normal by up to about 5 dB. Even so, there is a range of 35 dB in the mean speech voltage.

To handle such a total range of mean level and still allow sufficient dynamic range to cope with the peaks in speech requires the use of some kind of amplitude control. This may be voice operated gain adjusting devices (VOGADS) or companding.

Whilst VOGADS offer great adjustment range, they may introduce intelligible cross-talk from other telephone circuits and usually suffer from undesirable effects when used on circuits with noise between speech bursts, delayed echoes or when transients cause unwanted gain reduction. Further development may lessen these imperfections.

Most cellular systems have adopted companding, which provides a worthwhile improvement in audio signal-to-noise ratio and is less affected by noise and transients. At the transmitter the audio is compressed so that, for example, a 2 dB change in level is reduced to 1 dB. At the receiver the signals are expanded again. To minimize distortion caused by error in tracking between the compressor and the expander companding is usually limited to a 2:1 ratio, whilst 4:1, if attainable, would be useful in coping with the dynamic range of signals found in practice. The time constants of the companding functions are selected so that they work at a syllabic rate; peaks of speech within syllables are not companded.

Other factors to be considered in the transmission planning of cellular systems are stability and delay requirements. Whilst the radio network is usually 4-wire throughout, providing separate go and return paths, it has to connect to a telephone network that is 2-wire at some point. Miss-matching at the 2-wire to 4-wire conversion point can introduce a significant echo, which can cause difficulty in conversation, especially if delayed. Loss within the radio network, including between the transducers of the mobile handset, must be adjusted to ensure stability when connected to the telephone network. This limits the gain that may be applied in the cellular network to overcome circuit losses within the land telephone network.

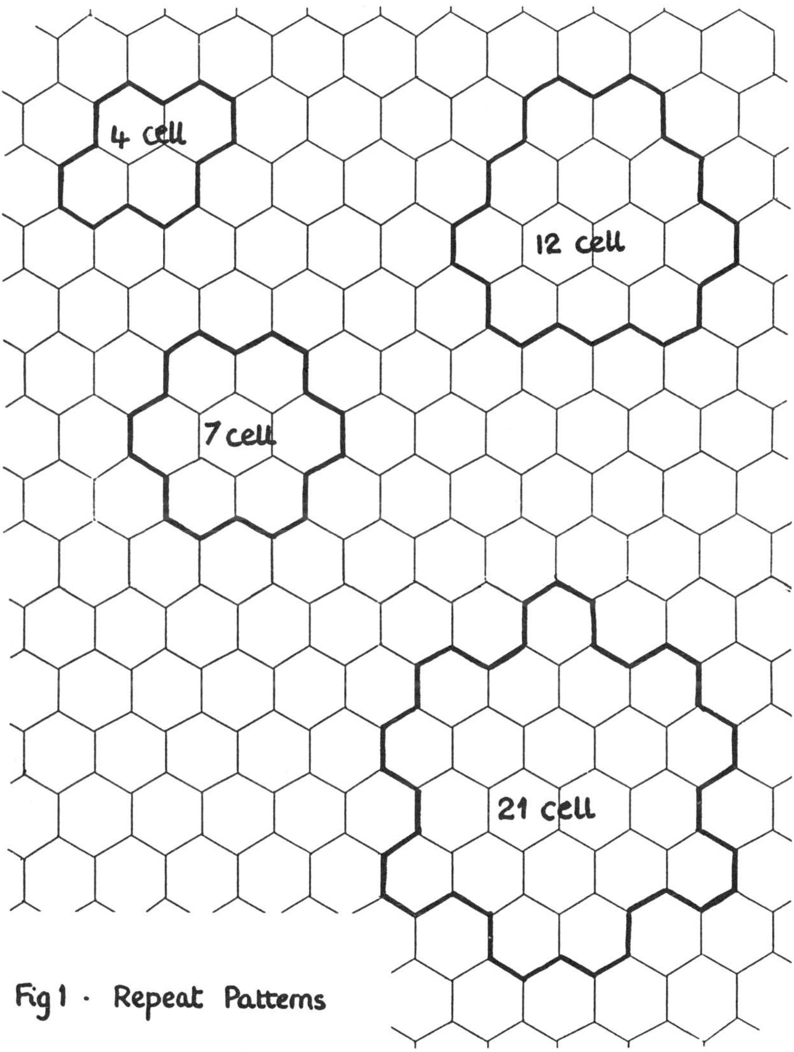

Fig 1 · Repeat Patterns

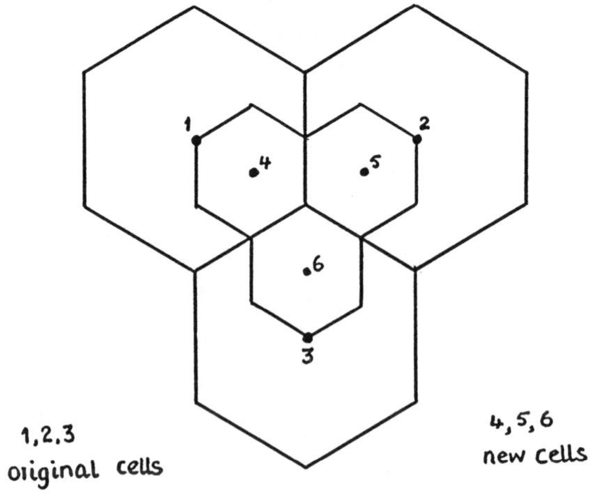

Fig 2. Additional Cells for growth

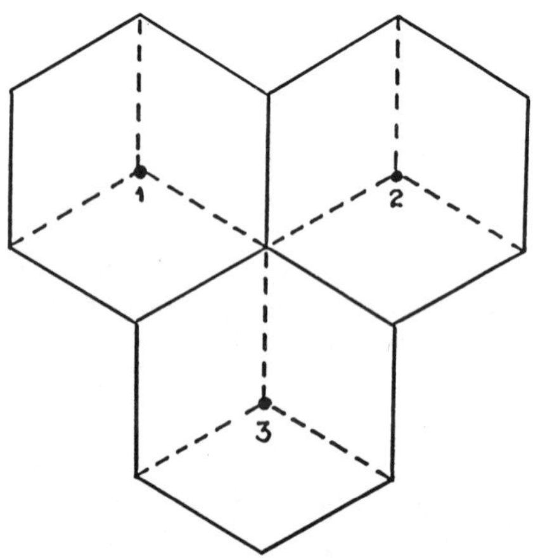

Fig 3. 120° Sector Cells for growth

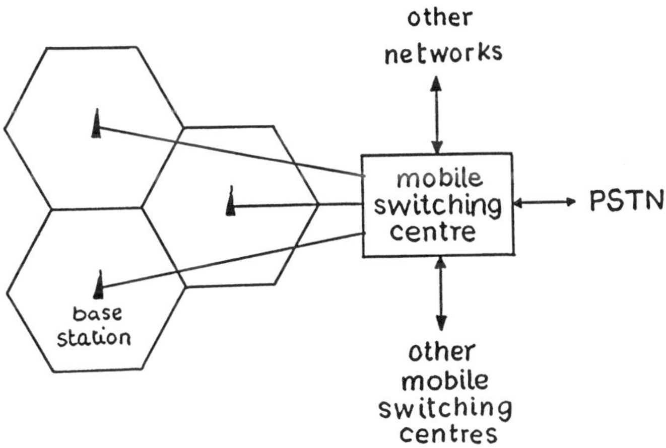

Fig 4. Network arrangement

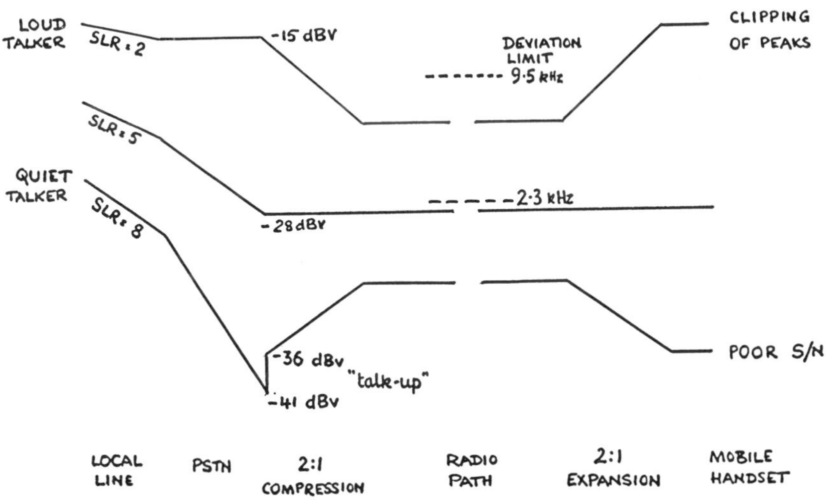

Fig 5. Transmission Diagram

Chapter 14

Mobile radio—the future

P.G. Farrell

14.1 INTRODUCTION

This chapter briefly summarises the material presented in the previous chapters, picking out those points which are of particular interest and relevance to the future development of mobile radio. In addition, certain topics not covered in the rest of the book, but which are important in the context of land mobile radio systems, are introduced.

14.2 MAIN THEMES

Chapters 1 and 2 described VHF and UHF radio propagation, particularly in urban areas. Various factors modify the flat earth inverse fourth-power relationship between path loss and distance from the transmitter, including shadowing, multipath fading, roughness of terrain, and urbanisation factor. Prediction models have been developed which compute median path losses fairly accurately, and a realistic national propagation map is beginning to emerge. It is worth commenting that field strength measurements and predictions should not be matters for commerical contention or secrecy, but should be held centrally and made easily available to all who need to use them for design purposes. Multipath models also exist, and though they suffer from certain shortcomings, they can be used to estimate fading (level crossing) rates, and average fade duration. These characteristics can then be used to make performance predictions for various types of modulation, diversity, coding, etc. These predictions are best used on a relative basis, to compare different modulation methods, for example, rather than as absolute measures of performance. Field measurements indicate that signals received with antennas about $\lambda/2$ or more apart (where λ is the wavelength of the carrier) are essentially uncorrelated; this enables the use of space diversity on mobiles, which gives a substantial increase in effective received signal power. It may also be possible to take advantage of field diversity, if the E and H fields fade independently.

Aerial and base station design aspects were covered in Chapter 4. The main problems which face designers arise from the need to cope with unwanted interfering emissions. These emissions are generated by coupling effects between site equipments, and by non-linearities in transmitting and receiving components or in metal structures (e.g., antenna masts) which re-radiate (the "rusty bolt" effect). The problems are greatly exacerbated at sites where several transmitters and receivers are co-located, as in practice is often necessary. This imposes severe constraints on components such as isolators, filters, feeders, etc. Very efficient hybrid combiners are also required, particularly for close frequencies, and Butler matrix network combiners appear to offer an effective solution to this problem. Even with the best of components, however, it is often not possible to neutralise all significant unwanted spectral emissions, especially interfering intermodulation products. The presence of these products, as shown in Chapter 4, severely restricts possible sets of channel frequency assignments. Because of the need to avoid channels containing intermodulation products, usable frequencies need to be relatively widely spaced. Thus third-harmonic-free minimum bandspread assignments for 3 channels are only 50% efficient, for 8 channels 20%, and for 13 channels 10% (the total number of channels, c_t, has an approximately square-law relationship to the number of useable channels, c_u: $c_t \approx (c_u-1)^2$). Determining the number and amplitude of the intermodulation products is a complex analytical problem; since non-linear reactive devices are involved, it is not possible to ignore phase relationships, for example. Much further theoretical and practical development is needed. Additional difficulties are caused by lack of co-ordination between users (e.g., in siting aerials) and poor maintainance (e.g., resulting in droopy (in the non-technical sense!) masts, booms and aerials).

The topic of modulation was dealt with in Chapters 5 and 6. This is one of the most controversial topics in mobile radio, including as it does the arguments over FM versus SSB, and analogue versus digital! A modulation method must be capable of conveying speech (still by far the most important service), encrypted speech, data, facsimile, and a number of other services (e.g., squelch enhancement, selective calling, message paging, etc.) To be efficient, to match the predicted growth in mobile radio traffic, and to facilitate the interface with the PSTN (and later the ISDN), these services should be conveyed in as integrated a way as possible (avoiding the use of data modems, for example). To be suitable, however, a method of modulation must meet a number of constraints. The chief of these is bandwidth: in the future the pressure to be bandwidth efficient will be inexorable, because of the demand for channels. Other constraints on a modulation method are that it:

be robust against fading, multi-path, and adjacent and co-channel interference;

degrade gracefully as the signal to noise and distortion (SINAD) ratio decreases;

provide signal level control and synchronisation;

be relatively simple to implement;

be power efficient;

lead to effective area coverage; and

provide the required grade of service (i.e., in terms of SINAD, error rate or error-free seconds).

AM and FM are well-established modulation methods for mobile radio, but neither (particularly AM) has the characteristics required to meet all the constraints noted above. This leads to interest (as described in Chapter 6) in certain other methods, including DSB and DSB-CC (controlled carrier), arc-sinc modulation (ASM) and the family of SSB, ISB (independent sideband) and carrier-position-modulation (CPM) or phase-locked SSB (PL-SSB) methods. These latter techniques are the most bandwidth efficient, but have the disadvantage of being relatively complex to implement. This disadvantage will be offset in the future, however, by advances in technology, and by the pressure to use digital modulation techniques. Digital modulation offers considerable advantages, notably:

the ability to provide a truly integrated service;

the better performance of regeneration over amplification;

the ability to take full advantage of digital signal processing, coding, switching and control techniques;

better coverage and range; and

higher link capacity and system throughput because of faster speed of transaction.

Digital modulation has some disadvantges; it is more complex and therefore initially more costly than analogue modulation, and it suffers from a bandwidth penalty. Both of these will cease to be significant problems as mobile radio expands and bandwidth efficient digital modulation schemes are further developed. Schemes of interest at the moment include quadrature PSK(QPSK), QAM, fast FSK(FFSK) or minimum shift keying (MSK), tamed frequency modulation (TFM), Gaussian MSK(GMSK), correlative PSK(COR-PSK), and others [1]. It is not necessary to use truly constant envelope modulation, as is often advocated; near-constant envelope schemes can be just as effective, and have less stringent requirements. Advances in speech coding will also help to make digital speech transmission more spectrally efficient [2].

Area coverage schemes, introduced in Chapter 7, are required when it is necessary to communicate with mobiles over an area that can only be covered by several base stations. Each base station may simultaneously transmit (or receive) the same signal on a different frequency (multi-frequency schemes); or, using the same frequency, only the nearest station communicates with the mobile (basic co-channel schemes); or all stations may simultaneously transmit the same signal but with carriers offset by several kHz (spaced carrier co-channel schemes). The first and third of these schemes are spectrally inefficient; the second (a form of selective calling system) is more complicated than the other two, but is the basis for operation of the cellular schemes described in Chapter 13. Quasi-synchronous operation maintains the spectral advantage of a basic co-channel scheme, without its complexity, using carriers offset by only a few Hz. Quasi-synchronous is most effective with AM and SSB modulation; sideband diversity much improves data transmission performance in overlap areas. The base station can also take advantage of return path diversity. Synchronous schemes (no deliberate carrier offsets) can be implemented with non-demodulating repeaters, but require radio links between the control point and the transmitting sites. Their performance is at least as good as that of quasi-sync schemes, and is better for data transmission and with FM modulation. A possibility for the future is to use on-frequency repeaters, which have been successfully operated at HF frequencies. Satellite mobile radio schemes are also being studied and developed [13].

Data systems for mobile radio were described in Chapter 8. The advantages of digital systems were mentioned above, but it is appropriate to re-emphasise here that a 3dB improvement in signal-to-noise ratio, brought about by the use of a digital modulation or coding scheme, is worth 20% extra range, and 40% extra area coverage. As shown in Chapter 8, it is best to use direct modulation if possible, and to avoid transmitting in the 100-500 baud rate range (i.e., 100-500 bit/sec if a binary system). Better performance is achieved by either sending at a slower rate, or at a faster rate with an error-detection-and-correction (EDC) code. In a quasi-sync scheme, the error rate experienced by a stationary mobile can be improved slightly by increasing the carrier offset, though this must be traded against the fading performance of a moving mobile. As mentioned above, it is possible to predict average error rates, and to very roughly estimate burst lengths (from fade duration predictions). Because errors occur in a diffuse mixture of burst and random patterns on mobile radio channels, these parameters alone are not sufficient to enable the design of a really effective EDC scheme. What is required is some knowledge of the error pattern characteristics, and the existence of a set of suitable diffuse and burst error control codes. Even if some knowledge of the former could be obtained (from channel measurements or more refined channel models), the latter would prove to be a problem, since this type of EDC code is not well developed at present. The best current solution is therefore to use random EDC codes with a sufficient degree of interleaving [3]; in the future, however, more efficient burst and diffuse codes are likely to emerge. Block codes are preferable to convolutional codes [4] if messages are short. As an alternative to forward error control (FEC), retransmission error control (REC) can be used; this is simple to implement because only error detection decoding is required, together with a protocol for requesting a repeat (ED-ARQ). REC is most effective when combined with a modicum of FEC, as in a hybrid FEC-ARQ scheme [4]. With careful EDC design, output (decoded) error rates in the range 10^{-4} to 10^{-6} are achievable with moderate system complexity. As simpler decoding algorithms are devised, and VLSI technology develops, better performance will become feasible in the future.

In one sense, paging systems (Chapter 9) have always been digital, since very few distinct messages are transmitted, and it is often only one: "you are wanted"! Thus it is unfair to say that digital pagers compare unfavourably with tone pagers; the decreased sensitivity of the digital pager is a result of increased paging system capacity, and the comparison is a false one. Inductive loop and leaky feeder transmission is widely used; this has the advantage of conserving spectrum, but is not suitable for wide area paging, where radio has to be used. The recently introduced UK national paging system is quasi-synchronous, with nominally zero frequency offsets, and direct conversion receivers. This system has the distinction of using a nationally agreed transmission format, the POCSAG code, which is a considerable success story for standardisation initiatives. A more dubious characteristic of any paging system, which it shares with portable radio and cassettes, and with digital watches, is its potential for creating noise pollution. This puts a constraint on paging system designers and users alike, to keep audible nuisance to a minimum. In the future, paging receivers are likely to become smaller and cheaper, putting them in the watch and calculator category. Paging signals may ride "piggy back" on broadcast transmissions, and are likely to be integrated into cellular mobile radio and cordless telephone systems, perhaps forming part of a personal radio system.

Regulatory control (Chapter 10) is another controversial aspect of mobile radio communications. The increasing demand for more mobile radio channels and for more advanced mobile radio systems, has placed considerable pressure on the regulatory authorities, and their response has not always been above complaint. On the one hand, there is a clear need to regulate and control the use of a shared resource, the frequency spectrum, in such a way as to maximise the benefit for all. Also, it is often forgotten that national regulatory bodies are bound to implement the decisions of World Administrative Radio Conferences, which have the force of legal treaties between nations, however unpopular these decisions may be in certain cases or localities. On the other hand, it is possible to feel a certain sympathy for the way in which the Citizens Band (CB) radio community was treated, for example; and it may be that certain other sensitive issues have still to be settled to the satisfaction of all concerned. The recent Merriman Report is to be welcomed for its positive approach: regulatory authorities should be better resourced, should be adequately supported by scientific and engineering studies, should operate more openly, and should actively judge and manage system needs.

In return, mobile radio users should make their needs better known, should demand more, and should lobby effectively together. That users should act together is desirable anyway, so as to get the most out of mobile radio systems; for example, operating a common base station for users sharing the same channel(s) in an area.

Switching and signalling, covered in Chapter 11, will become more and more important as the mobile radio telephone service grows. A mobile telephone exchange (MTX) is an essential part of a mobile radio system, to interface it with the PSTN and, eventually, with the ISDN. Channel assignment in mobile radio is equivalent to FDM switching and concentration. The advantages of digital switching apply to an MTX just as much as to any other exchange. Effective signalling is vital to the efficient operation of an MTX, not only to establish and close down a channel before and after connection, but also to handle those conditions which are peculiar to an MTX, such as dealing with a signal which has disappeared because a mobile has entered a tunnel, or during hand-off between cells in a cellular system (Chapter 13), for example. Both in-band and common channel signalling is likely to be used in an advanced mobile radio system; the communication signals themselves will also be used to help control transmission, as in voice operated gain adjustment (VOGAD). The ultimate aim must be to completely integrate a wide area or national mobile radio system into the PSTN or PSDN, including a common numbering scheme. This last point implies that there must be international agreement on mobile numbering plans.

The basic aim of a radio trunking scheme (Chapter 12) is to mediate between 100% spectral efficiency, which means intolerably long queues for each frequency channel, and zero waiting time for a connection, which leads to very poor spectral utilisation. A trunking scheme can improve efficiency of use by 2-6 times, and is worth while even if applied to only three frequency channels, though the more channels that are trunked the greater the advantage. A trunking scheme must involve contention control (using a suitably efficient protocol), quick assignment of channels, queueing of calls (though not necessarily on a first-come-first-served basis), and rapid and very secure clear down. Unlike in wide-area or national mobile radio-telephone systems, private mobile radio systems do not usually interwork between themselves, or with the PSTN. By pooling the resources of several systems, however, more efficient operation and trunking is possible, and in the future the pressure to collaborate in this way will surely increase. Being part of an integrated mobile radio system provides these advantages anyway.

Many of the techniques suggested above for improving the efficiency and coverage of a mobile radio system, are incorporated in recently proposed and commissioned cellular schemes (Chapter 13). The area covered by a cellular base station is deliberately limited to a relatively small cell, so that the frequency channels used in the cell can be re-used in other (non-adjacent) cells, depending on the desired signal to interference ratio. By making the cells small enough, the capacity of the system can, in principle, be made as large as desired. In practice, capacity is limited by the number of frequency channels available to each cell, by the need to hand-off calls as mobiles cross cell boundaries, by the difficulty of establishing well-defined cells of the desired size, and by the complication of trunking, switching and controlling a very large number of small cells. Nevertheless, cellular radio is the only known scheme that can hope to meet the predicted demand for mobile services. Thus much effort is being put into research and development of cellular schemes.

14.3 ADDITIONAL TOPICS

This section introduces four additional topics that were not covered in the rest of the book. They are all techniques which are capable, actually or potentially, of improving the performance of a mobile radio system.

Soft decision detection is an example of integrated demodulation and decoding for digital modulation schemes with EDC coding [5]. Instead of making a hard decision on each digit received, a soft-decision detector first of all decides whether the signal is above or below the decision threshold, and then computes a "confidence" number which specifies how far from the threshold it is. This number is therefore a measure of how reliable the digit is; the reliability is taken into account in the decoding process which then follows. A soft-decision decoder operates with more information about the received signal than a hard-decision decoder does (e.g., it uses Euclidean instead of Hamming distance), so it has a performance approaching that of a matched filter for the code in use. The performance improvement, for channels perturbed by Gaussian additive white noise, is equivalent to about 2dB in signal-to-noise ratio; on fading channels like the mobile radio channel the improvement in coding gain is much greater. A soft-decision decoder is more complex than a hard-decision decoder, but not excessively so, particularly if used with array or cyclic EDC codes [4], and is much less complex than a full matched filter implementation.

The performance of a radio system demodulator/detector can be improved by combining, at the receiver, two or more versions of the same transmitted signal [6,7]. Space, time, field, polarisation, and sideband diversity are all of potential value for mobile radio systems, but not frequency diversity, because of the need for spectral efficiency. Possible techniques include switching diversity, pre-detection combining (e.g., equal gain, and ratio (Law assessor) combining), and post-detection combining (e.g., mode averaging, and confidence sum (soft-decision detection)). Diversity can provide 7-12dB of gain, or 1-2 orders of magnitude reduction in error rate. In addition, co- and adjacent-channel interference is reduced, and demodulation, coding, synchronisation, etc., is made easier. Disadvantages are the extra equipment required (particularly for post-detection combining), and the complexity of combining (signal assessment and co-phasing). Diversity is not always feasible for practical reasons (e.g., size, as in personal radios), but has been rather unjustly neglected in spite of the advantages it has to offer. Diversity is certainly likely to be used in future mobile radio systems.

Spread spectrum signalling is a form of frequency diversity. In spite of what was said in the previous paragraph against the use of frequency diversity, it turns out that spread-spectrum modulation can have a spectral efficiency at least as good as that of a conventional digital modulation scheme (8,9). Both direct-sequence and frequency-hopping techniques can be used to generate the spread-spectrum signals; frequency hopping cellular systems with spreading factors of 1000 or more can cater for millions of users. Spread-spectrum systems require very little synchronisation between users (sub-sets of users can act independently, for example), the system performance degrades gracefully as more and more users access the system (avoiding the sudden over-load characteristics of conventional systems), and roaming over a wide area is very easy to implement. There are a number of severe problems, however, particularly the difficulty of obtaining permission to use the very wide bandwidths required, and the cost and complexity of implementing matched filters for the large number of spread spectrum codes.

In some ways collaborative coding techniques [10,11] combine the advantages of spread-spectrum and conventional digital modulation methods. Collaborative coding offers the possibility of two or more users transmitting digital signals simultaneously in the same frequency channel. In this case, errors occur not only because of noise and incidental inter-ference, bit also because of the self-interference and contention effects between the users on the same channel. The code then has to resolve the contention between the two users. There is a wide range of these collaborative coding techniques [12], and their use is especially appropriate and potentially useful in mobile radio systems, where bandwidth is a very restricted resource. In theory, collaborative coding offers considerably increased mobile radio system capacity, with acceptable performance, at the expense of the more complex modems required to receive simultaneously transmitted signals. The use of this technique remains speculative at the time of writing, however; there are many research problems to solve before it can be implemented in a practical system.

14.4 CONCLUSIONS

The pressures which will shape the future of mobile radio are clear: the large demand for service, the need to be spectrally efficient, and the regulatory constraints. The implications of these pressures are that:-

(i) the demand must be responded to (nobody thought that the telephone would catch on to the extent that it has!);

(ii) spectrally efficient methods must be used (SSB/digital modulation, common base station, trunking, diversity, direct conversion, direct modulation, coding, efficient filtering, signal propagation and antennas, etc.);

(iii) cellular schemes are unavoidable, perhaps in combination with spread-spectrum (or collaborative coding) methods;

(iv) the signal must be got where it is wanted: micro-cellular, satellite, and even balloon mobile radio systems are worthy of consideration;

(v) users must get together: in the future mobile radio will be an integrated digital system combining radio-telephone, cordless telephone, personal radio, paging, etc., with transparent interfaces to the ISDN.

REFERENCES

1. Clark, A. P., "Digital modems for land mobile radio", to be published in a forthcoming special issue of IEE Proc. Part F on mobile radio systems.

2. Darnell, M., "Speech digitisation techniques", IEE Conf. on "Secure Communication Systems", York, 22-23 Feb. 1984, Conf. Pub. No. 231, pp. 20-31.

3. Mabey, P. J., "Mobile radio data transmission - coding for error control, IEEE Trans. Veh. Tech., Vol. VT-27, pp. 99-109, Aug. 1978.

4. Farrell, P. G., "A survey of error-control codes", in "Algebraic Coding Theory and Applications", ed. G. Longo, Springer-Verlag, 1979.

5. Farrell, P. G. and Goodman, R. M. F., "Soft-decision error control for HF data transmission", Proc. IEE, Vol. 127, Part F, No. 5, Oct. 1980, pp. 389-400.

6. Parsons, J. D. and Pongsupaht, A., "Error-rate reduction in VHF mobile radio data systems using specific diversity reception techniques", IEE Proc. Vol. 127, Part F, No. 6, Dec. 1980, p. 475.

7. Henry, P. S. and Glance, B. S., "A new approach to high-capacity digital mobile radio, Bell Syst. Tech. Jour., Vol. 60, No. 8, Oct. 1981, pp. 1891-1904.

8. Yue, O.-C., "Spread-spectrum mobile radio, 1977-1982", IEEE Trans. Veh. Tech., Vol. VT-32, No. 1, Feb. 1983, p. 98.

9. Buckingham, S. D., "The spread-spectrum controversy", Elec. and Power, March 1980, pp. 222-224.

10. Farrell, P. G., "Frequency sharing by means of collaborative coding techniques", IEE Conf. on "Radio Spectrum Conservation Techniques", Birmingham, 6-8 Sept. 1983.

11. Farrell, P. G. and Abdul-Reda, A. J. A., "Performance of collaborative coding schemes for frequency sharing in mobile radio", IEE Conf. on "Mobile Radio Systems and Techniques", York, 10-13 Sept. 1984, Conf. Pub. 238.

12. Farrell, P. G. "Survey of channel coding for multi-user systems", in "New Concepts in Multi-User Communications", Ed. J. K. Skwirzynski, Sijthoff and Noordhoff, 1981, pp. 133-159.

13. Watson, P. A., Gardiner, J. G. and Seneviratne, A. P., "Proposal for a land mobile satellite communications experiment", IEE Conf. on "Mobile Radio Systems and techniques", York, 10-13 Sept. 1984, pp. 126-130.

Index